ANNALS OF
THE NEW YORK ACADEMY
OF SCIENCES

Volume 412

EDITORIAL STAFF
Executive Editor
BILL BOLAND
Managing Editor
JOYCE HITCHCOCK

The New York Academy of Sciences
2 East 63rd Street
New York, New York 10021

HISTORY AND PHILOSOPHY OF SCIENCE: SELECTED PAPERS

ANNALS OF THE NEW YORK ACADEMY OF SCIENCES
Volume 412

HISTORY AND PHILOSOPHY OF SCIENCE: SELECTED PAPERS

Edited by Joseph W. Dauben and Virginia Staudt Sexton

The New York Academy of Sciences
New York, New York
1983

Library of Congress Cataloging in Publication Data

Main entry under title:

History and philosophy of sciences.

(Annals of the New York Academy of Sciences; v. 412)
Selection of papers presented at monthly meetings of the Section for History, Philosophy, and Ethical Issues of Science and Technology of the New York Academy of Sciences between 1979 and 1981.
Bibliography: p.
Includes index.
1. Science—History—Addresses, essays, lectures.
2. Science—Philosophy—Addresses, essays, lectures.
I. Dauben, Joseph Warren, 1944– . II. Sexton,
Virginia Staudt. III. New York Academy of Sciences.
Section for History, Philosophy, and Ethical Issues of
Science and Technology. IV. Series.
Q11.N5 vol. 412 [Q126.8]   500 [509]   83-19383
ISBN 0-89766-217-2
ISBN 0-89766-218-0 (pbk.)

SP
Printed in the United States of America
ISBN 0-89766-217-2 (Cloth)
ISBN 0-89766-218-0 (Paper)

ANNALS OF THE NEW YORK ACADEMY OF SCIENCES

VOLUME 412

October 14, 1983

HISTORY AND PHILOSOPHY OF SCIENCE: SELECTED PAPERS[a]

Editors

JOSEPH W. DAUBEN and VIRGINIA STAUDT SEXTON

CONTENTS

[a]The papers in this volume were presented at meetings of the Section of History, Philosophy, and Ethical Issues of Science and Technology of The New York Academy of Sciences during the years 1979–1981.

Preface

JOSEPH W. DAUBEN[a]

Department of History
Herbert H. Lehman College
The City University of New York
Bronx, New York 10468

This volume of essays represents a selection of papers presented at monthly meetings of the Section of History, Philosophy, and Ethical Issues of Science and Technology of the New York Academy of Sciences between 1979 and 1981. During this period Professor Virginia Staudt Sexton chaired the Section's activities, and the works published here reflect the diversity of subjects taken up by the Section under her direction. Unfortunately, it was not possible to include in this *Annal* all of the papers actually presented at the Academy, and mention should at least be made of two that drew audiences in record numbers: Frank J. Sulloway (Massachusetts Institute of Technology) presented a lecture on *Freud and Biology: The Hidden Legacy,* and Barbara Gutmann Rosenkrantz (Harvard University) at the last meeting in May of 1981 discussed the history of mental health care in a paper on *Deinstitutionalization: A Hundred Year Perspective.*

Despite the fact that it was impossible to include all of the papers presented at the Academy during the two years Professor Sexton served as Chairman of the Section of History, Philosophy and Ethical Issues of Science and Technology, those included here clearly demonstrate the success of the Section in bringing distinguished scholars to New York, and making available their research to members of the New York Academy of Sciences, as well as to a broader general audience drawn from the greater Metropolitan New York area.

[a]Chairman, 1981–1983 of the Section of History, Philosophy, and Ethical Issues of Science and Technology.

Medical Care in the Countryside near Paris, 1800–1914[a,b]

EVELYN ACKERMAN

Department of History
The City University of New York
City University of New York
Bronx, New York 10468

At the beginning of the nineteenth century, the Napoleonic government took steps to spread medical care to rural areas of France. A circular of May 2, 1805 from the Minister of the Interior to all the prefects of the empire instructed them to appoint a doctor to be in charge of epidemics in each *arrondissement*.[1] The position of epidemic doctor had its origins in the eighteenth century when the intendants of the various provinces had occasionally sent physicians to areas heavily stricken by disease, but the nineteenth-century epidemic doctor was to have a more important and less intermittent role.[2]

The epidemic doctors were to function as a link in a chain of medical police, and I use the expression "medical police" advisedly, for this is just about the time when Johann Peter Frank's writings on that subject appeared.[3] It was the responsibility of the mayor of each town to keep a reasonably vigilant eye on the health of his commune, but as soon as as epidemic was suspected, he was to notify the subprefect of the arrondissement who would dispatch the epidemic doctor.

Once arrived on the scene, the epidemic doctor had several duties. Naturally, he was to visit the sick, prescribe medicines and perform procedures, and order, if necessary, the free distribution of bouillon, meat, and wine. If there were a local doctor or health officer,[4] the epidemic doctor was to advise and possibly instruct him in the proper treatment of the disease. Gastrointestinal disorders such as typhoid, malarial-type fevers, and various forms of dysentery were the most frequently encountered illnesses. Cholera struck France several times during the nineteenth century, in 1832, 1849, 1854, and 1866.

But the epidemic doctor's duties were not limited to the simple management of symptoms. A second circular from the Minister of the Interior in 1816 makes it clear that the epidemic doctor was to be a soldier in the battle to civilize the peasants. He was to help the local administration in preventing mass panics during outbreaks of disease; he was to enlighten rural people

[a]This publication was supported in part by National Institutes of Health Grant LM 03001 from the National Library of Medicine. Support was also provided by the Research Foundation of the City University of New York.
[b]This paper was presented at the September 26, 1979 meeting of the Section of History, Philosophy and Ethical Issues of Science and Technology of the New York Academy of Sciences.

1

about "personal hygiene [which is] much too neglected in the countryside"; he was to try to do something about the eating habits of the peasantry who "often exist only on a coarse diet."[5] In short, he was to try to change the patterns of behavior that rural people had evolved to deal with the problems of scarcity, fear, and uncertainty that were a large part of peasant life, at least before 1850.

This paper will discuss the functioning of this rudimentary public health service, and subsequent public health and medical services in rural France between the time of Napoleon and the First World War. Of special concern will be popular reaction to these efforts to spread health care. The geographical focus will be one French department, the department of the Seine-et-Oise. Located right next to Paris, the Seine-et-Oise had a population of 433,000 according to the census taken in 1806.[6] With the exception of its capital city, Versailles, which numbered 27,000 people, the department was overwhelmingly rural. According to a list of medical personnel compiled in 1809, the Seine-et-Oise boasted thirty-nine physicians, eleven of whom practiced in the city of Versailles.[7] An additional eighty-seven health officers were active in the department, and if we group physicians and health officers together as qualified medical practitioners, then we come up with a figure of one medical practitioner in 1809 per 3400 people when Versailles is included, one per 3800 when it is not.

Clearly, then, in the eyes of the administration at least, there was room for improvement in the supplying of health care in the Seine-et-Oise; the prefect quickly named five epidemic doctors, one per arrondissement, within several months of the Minister of the Interior's circular of 1805. In 1827, the prefect expanded the epidemic service by adding an epidemic doctor for each canton, and so, by the close of the Bourbon Restoration, the public health corps of the Seine-et-Oise numbered slightly over thirty people, mostly physicians but also, especially in the outlying cantons, some health officers.[8]

What followed from the creation of this epidemic service was something more than the orderly distribution of smallpox vaccination and quinine; it was a veritable clash between two opposing cultures with different norms, styles, and expectations. For the physicians of the Seine-et-Oise, rural people were dirty and stubborn, vulnerable children who needed guidance and direction. Mention of the filth of rural dwellings and villages appeared in virtually every report the doctors sent in to the subprefects; laments about peasant reluctance to seek out medical help were also legion. In 1802, Dr. Engaz, a surgeon who treated an epidemic at Maisse, noted that "few people had recourse to the resources of medicine which were an effective aid for those who did,"[9] and in 1839, Dr. Reybaud, the epidemic doctor for the canton of Milly, cited several deaths at Moigny as being due to "the deplorable negligence of the sick most of whom either do not ask for medical care until it is too late, or do not ask for it at all."[10]

It is hard to see any change in this attitude before the Revolution of 1848. The occasional testimonies that indicate a greater willingness to accept medical care are often puzzling. In 1842, the mayor of the fairly prosperous town of Ville d'Avray, writing about a measles epidemic, noted that "the child who died and the one who is still in danger belong to a locksmith and a quarry worker, who are fairly well off and who spared no expense in giving them all necessary care." But these families were apparently the exception; one way the town dealt with the epidemic was to have the doctor from the local hospital come every day to provide free medical attention for children with measles. "In this way," continued the mayor, "we avoid that any family, by cruel parsimony, get the idea of not calling the doctor, or of calling him when it is too late."[11]

The physicians not only saw the peasants as stubborn; they (correctly) saw them as weakened by their poverty. Dr. Lelarge, the epidemic doctor of the arrondissement of Pontoise, analyzing an 1807 epidemic of intermittent fever in Gagny, eloquently described the circular problem of the poor. Exhausted by disease, itself greatly aggravated by their chronic deprivation, the people of Gagny were too weak to do their usual work. Wrote Dr. Lelarge:

> Their pale, powdery complexions, their hollow cheeks, their swollen legs are all so many indications of the frightful poverty into which they have fallen. The filth of their houses, the dirt with which they are covered, their tattered clothing, the scarcity of furniture, the black bread they eat, form the most heartbreaking sight. Well taught and especially given effective aid, they would doubtless quickly emerge from the listless state into which they are plunged.[12]

And even some thirty years later, in 1839, when references to abject poverty were becoming a less inevitable part of each report, Dr. Bourgeois, the epidemic doctor of the arrondissement of Etampes, could write of an outbreak of intermittent fever at Gironville that

> most of the sick are either in poverty or have only the strength of their arms with which to make a living; most of these sick people begin working as soon as they have barely recovered, and relapse, because they cannot continue the very expensive treatment that they need long enough.[13]

Economic deprivation was not the only factor that weakened the peasants; the doctors felt that they were more susceptible to disease because, like children, they were largely incapable of regulating and controlling their emotions. Although the pathogenic role of strong feelings was a commonplace of early nineteenth-century medicine, the physicians of the Seine-et-Oise placed a very strong emphasis on its role in exacerbating peasant ills. The subprefect of Pontoise summed up this position very well in 1821 in a letter to the prefect of the Seine-et-Oise in which he pointed to "the fear that seizes imaginations as feeble as those of most country dwellers as being "usually the most lethal circumstance of these sorts of actions."[14]

Nor were the peasants disciplined enough to regulate their appetites, according to the doctors who repeatedly deplored the peasants' dietary excesses. Many ills, from simple gastroenteritis to cholera, were blamed on impulsive indulgence in heavy foods like blood sausage and hardboiled eggs, not to mention alcohol. Learned ideas about self-control, however, had little place in the way of life that peasants had developed to deal with the main features and stresses of their existence. Rural life, especially in the opening years of the nineteenth century, was still punctuated by dramatic variations in food supply. The disastrous harvest of 1817, for example, was as bad as any of the eighteenth century, at least for the Seine-et-Oise where, as in most parts of northern France, popular diet was heavily dependent on bread. Village life was dotted with such rituals as the killing of the pig in late November, and by a period of serious overeating in the days that followed. Wedding feasts could last for a day or two, and chronically malnourished people would suddenly overload their stomachs with unaccustomedly rich food.

There were other reasons, too, for peasant reluctance to consult physicians, or to accept their ministrations. Although French doctors fought energetically in the first half of the nineteenth century to obtain a monopoly of the healing business, they had little real scientific or technical knowledge with which to justify these pretensions. Popular healers were more familiar figures, less threatening, less condemning. And before 1850, they were not necessarily less effective in combatting disease. Physicians really only had two fairly sure treatments for specific diseases. One of them, quinine, was readily accepted by rural people because of its clear effectiveness in combating malarial-type fevers. Dr. Texier, the epidemic doctor for the arrondissement of Versailles, reporting on an outbreak of "fevers" in Gif in 1811, observed that "in the beginning, several sick people refused to take this medication because of fairly ridiculous or ill-advised notions they got from [local] surgeons, but after seeing the excellent effects it had on their neighbors, everyone hurried to ask for it."[15] But the second useful medication, the Jennerian vaccination against smallpox, which was introduced in the Seine-et-Oise in 1801, met with severe resistance from a population unwilling to allow itself to be injected with a foreign substance.

Indeed, quinine was the only remedy people sought out. The usual battery of techniques employed by the doctors of the Seine-et-Oise featured many uncomfortable and frightening treatment methods. Charles Rosenberg, in an excellent article in *Perspectives in Biology and Medicine,* has shown how a dramatic technique—a violent purge, a painful blistering agent—was part of a cultural ritual, part of a curing process that the patient expected and would have felt deprived of had it not been there.[16] Still and all, there is a fine line between the dramatic and the disgusting, and many of the methods used by the Seine-et-Oise physicians easily fell into both categories. Even for people accustomed to bleeding and purging, for example, Dr. Fossoyeux's method of handling the sweating sickness that occasionally followed cholera must have seemed strange and punitive. Sweating sickness, as its name implies, involved profuse perspiration and also extreme thirst. For the thirst, citrus drinks, reasonably enough, were prescribed, but, bewilderingly, the sweating was managed by refusing to let the patient change his soaked bedclothes. Predictably, patients were frequently uncooperative, and here is how Dr. Fossoyeux, the epidemic doctor for the canton of Ecouen, dealt with their objections:

> When they complained of excessive heat, of pain in the thorax, of difficult breathing and frequent palpitations, of anxiety, or chest pains and especially of

that overwhelming restlessness that made them want constantly to change their linen, their position, even their bed . . . , [this] seemed to indicate the necessity of a bleeding to prevent some sort of congestion. I did not hesitate to open a vein once or twice, and I often had success.[17]

From Dr. Fossoyeux's theoretical standpoint, this procedure could be justified in the sense that blood was diverted from a site of excitement, and further excitement was prevented. But from the patient's point of view, bleeding at such a time may well have appeared like a punishment for his understandable agitation about being constrained to remain in sweaty bedclothes.

With this background of mutual suspicion between the medical corps and the peasantry, it was almost inevitable that the cholera epidemic in 1832 would be a time of unmitigated panic. No other disease gripped popular imagination with as much fear during the nineteenth century as cholera. The onset of cholera was terribly dramatic. The victim, very often a person in the prime of life, would be struck by violent diarrhea and cramps. Within a few hours, he would become dehydrated and shriveled; his skin, especially his face, would blacken; he would become very cold and die in between 25 and 50% of all cases.[18] Cholera was a disease, then, that would look remarkably like a direct punishment from God in retaliation for a very specific sin. For all these reasons, cholera was to elicit a surge of panic that no disease had produced in Europe since the last outbreaks of the plague at the end of the seventeenth century.

Cholera is a waterborne disease transmitted by the cholera vibrion. When the excreta of a person affected with cholera enter the drinking water of a cholera-free population, a certain number of people in the previously healthy population will develop the disease. Cholera flourishes in situations of general filth and poor sanitary conditions. The villages of the Seine-et-Oise, then, with their nonexistent plumbing, with their manure piles at peasants' doors, were obviously extremely fertile soil for the disease. Between April and October 1832, roughly twenty thousand cases of cholera were reported in the department of the Seine-et-Oise.

The administration and the medical corps of the Seine-et-Oise fought cholera energetically, but with little real skill. There was absolutely no agreement among doctors about therapeutic procedure; indeed, the 1832 epidemic seemed to bring out professional rivalries among the different doctors. Dr. Dussaux of Mantes, for example, roughly attacked the "humdrum and empirical"[19] treatment methods used by Dr. Giard in the village of Issou near Mantes. Dr. Fossoyeux from Ecouen, whom we have already met, deplored even more than competition from local wise women "the pride, the jealousy, the blindness, the badly directed zeal, and the obstinacy of [his] predecessor."[20] Charlatans and amateur healers did a record business during the cholera epidemic. Reports of the activities of these entrepreneurs poured in from all parts of the department; in the arrondissement of Pontoise, it was country women who freely gave advice; in the arrondissement of Corbeil, Dr. Petit complained of "a distributor of preserving water and curing liquor [who] has developed such a following that this man, a former army officer with limited mental capacities, appears quite impressed with his own importance."[21]

Popular willingness to seek out medical care, lukewarm in even the best of times, all but disappeared. Hostility toward physicians took its most dramatic form in a series of accusations that the doctors of 1832 had purposely infected people with cholera, had poisoned them; sometimes the rumor had it that they had been paid by the central government to do so. Dr. Dussaux of Mantes noted these accusations "not only among the poorest and most ignorant classes in the countryside, but also among men whose social position and upbringing should shelter them from such stupidity."[22] Dr. Peyron of Marines observed that an ad hoc pharmacy set up in the village of Chars was not used by the townspeople "because most [of them] saw in this public interest measure only an abominable conspiracy of the authorities and the doctors against the health of the people of Chars."[23]

When cholera returned to the department seventeen years later, popular and administrative reaction to the second epidemic differed from reaction to the first in two ways. While it would be naïve to assert that the people of the Seine-et-Oise stood by calmly in 1849 as cholera spread, the elements of wild panic that were present in 1832 seem to have been rarer. Dr. Lemazurier, who had tended the population of Versailles during both epidemics, noted this change, as did his colleague Dr. Bourgeois of the arrondissement of Etampes, and the Hygiene Council of the arrondissement of Corbeil. The general feeling was expressed succinctly by the Hygiene Council in its report of 1850 to the Prefect:

> Popular attitudes in each of the places affected by cholera in 1849 demonstrated that notable progress had taken place in the mentality and habits of country people and city dwellers alike. Cholera doubtless caused real fear and terror . . . , but we did not see, as was the case in 1832, unreasoning crowds scream about poisonings, turn like wild animals on the very people who tried to help them . . . ; people finally realized that the disease was an egalitarian evil, striking its victims randomly.[24]

The second change in the way the 1849 cholera epidemic was handled was the one with the greater implications for the future. A decree of the executive power of the Second Republic of December 18, 1848 had ordered the establishment of hygiene committees in each arrondissement and in each department. By the end of 1849, these committees had been set up and had begun to function with considerable local support, especially from the educated classes. No one had great faith in physicians' ability to cure cholera by medical means, but increasingly, people began to respect and heed the admonitions of the same doctors and of administrators when they spoke about clearing stagnant water from village streets, removing manure piles from the doorsteps of peasant dwellings, cleaning up public latrines. It was in these areas, and not in the field of formal medicine as such, that the beginnings of progress were to be made in the Seine-et-Oise in the middle of the nineteenth century.

Let us turn now to the hygiene committees established in the Seine-et-Oise after 1849. Although councils were formed on the level of the arrondissement and even on the cantonal level in several cases, the Departmental Hygiene Council was the most aggressive one, planning and supervising activities for the entire department. Meeting in Versailles on the last Wednesday of every month, this committee, whose size ranged between ten and nineteen in the 1849–1914 period, included physicians, veterinarians, pharmacists, architects, engineers, and other local notables. At least two-fifths of the membership at any given time was physicians.

The work of the council between 1849 and 1914 splits easily into two periods, with the early 1880s, when medical practitioners began to understand certain disease processes better, as the dividing line. The insights of Pasteur and Koch were rather quickly translated into action on a public health level, as the Departmental Hygiene Council and physicians became enthusiastic supporters of various means of disinfection. Furthermore, after the 1880s, the Departmental Hygiene Council began to organize its activities along the lines of specific campaigns against clearly defined ills: unacceptably high infant mortality, especially among wet-nursed and bottle-fed babies, and tuberculosis. By contrast, it is clear from the council's efforts in the 1849–1880 period that these men saw themselves as pioneers; avid data-gatherers, they aimed to collect information on a myriad of subjects and to boil it down to a usable form from which to create a coherent picture of the public health life of the department; propagandizers, they planned to educate rural people about the need for cleanliness.

The question of some form of health education for the masses had been raised in the years surrounding the Revolution of 1848. During the July Monarchy and especially in the 1840s, as some doctors began to have more pretensions to social mindedness, several hygiene manuals appeared.

Physicians throughout France did not wholeheartedly endorse the notion of instructing peasants in hygiene. Some doctors were more comfortable with the earlier notion of the peasant as a child to be ordered about rather than as a reasoning person who could be educated. Amédée Latour, physician, medical journalist, and vociferous defender of physicians' interests, wrote such a scathing review of one of these hygiene manuals in his newpaper *L'Union médicale* that the authors of the manual were compelled to respond. Latour had argued that to teach peasants that their living conditions could be improved would incite revolutionary feelings. To which the authors countered, "What, a poor man cannot leave his foul, damp hovel for a drier, less pernicious abode, he cannot launder his clothes in the river, or wash his hands and face regularly without threatening the very basis of society?"[25]

The doctors on the Departmental Hygiene Council of the Seine-et-Oise would have seconded these sentiments. In their meeting of June 1849, they had only high praise for a hygiene manual prepared for schoolchildren by Amédée Eugène Bataille. Dr. Bataille, a member of the council, was both a physician and a pharmacist; his book, *Precepts of Hygiene for Schoolchildren,* contained ten chapters on health- and hygiene-related topics for people, and six chapters on the proper care of animals. In 1852, the Medical Society of the Arrondissement of Rambouillet asked the prefect for a subsidy to print more copies of Bataille's book; citing "the precision and practical wisdom given to the poor class of farmers," the prefect made the funds available.[26]

Mayors throughout the department encouraged distribution of the manual in both boys' and girls' schools.[27]

During the 1850s, Louis Napoleon, Emperor of the French, encouraged his prefects to set up and fund on the local level free medical care services for the needy poor. In the Seine-et-Oise, the Departmental Hygiene Council quickly became involved with the administration of this service, which included the vaccination program as well. Started in 1854 under the watchful eyes of the medical corps, the free medical care service in the Seine-et-Oise grew and flourished until 1893, when a national law ordered the creation of these services throughout France. Instituted at least in part to create work for the overcrowded medical profession, the free medical care service helped medicalize the rural population by accustoming about 15 percent of it to the ritual of an annual visit to a doctor.

As we study the records of the Departmental Hygiene Council, the reports of the doctors who participated in the free medical care service, and the correspondence between the epidemic doctors and the local administration, it is hard to avoid the impression that the task of all these advocates of health becomes easier after about 1870. When pro-health representatives speak to the peasants and to the mayors and town councils of rural villages, they can hardly be said to be preaching to the converted, but they are not crying in the wilderness either.

There are several reasons for this change from the early nineteenth century. For one thing, popular diet had improved markedly, both in quality and in quantity. This change was noticeable in the 1860s, and it continued until World War I. As many people deserted the countryside in the rural exodus, those members of the traditionally marginal class, the agricultural laborers, who remained on the land, were able to command better wages. At the same time, the nature of northern French agriculture changed. From the 1870s onward, the French market was flooded with cheap Russian, American, and Canadian wheat. Not only was the price of wheat driven down, thus assuring people of a steady bread supply, but French grain farmers began to raise more animals. With their higher wages, French agricultural laborers could now buy more meat; with the drastically improved wheat supply, they need not worry about obtaining bread. Agricultural surveys taken in the 1860–1880 period testify eloquently to the peasants' greatly improved diet.

At some point during the Third Republic, then, a good many people in the Seine-et-Oise began to change their diet from the heavily starchy one of their youth to a more well-balanced regime. The increased protein intake must have made them feel stronger and peppier, and this on a fairly continual basis. The important lesson that the people of the 1870s and 1880s were receiving, then, was that there were things you could do to your body—feed it better—that would result in your feeling more energetic. The subconscious message may well have been the beginnings of a belief that people could intervene to change the age-old inevitability of disease and of man's powerlessness in the face of it.

Literacy also had increased since the early nineteenth century, especially among women. In many of the less isolated towns in the Seine-et-Oise, male literacy, as measured however imperfectly by the ability to sign one's marriage certificate, had been as high as 75 percent at the turn of the nineteenth century; female literacy, however, had been woefully lower, closer to 50 percent. The Guizot Law of 1833 led to increased school attendance, and within a generation of its passage, female literacy in one representative town went from 73 to 93 percent.[28] If we can argue that exposure to school, book learning, and rational thinking favors a concern for health and hygiene, then the rise in female literacy is most significant, for it is the women in the family who make the decisions about health, especially children's health.

The war of 1870 also aided the transformation of mentalities by making reliance on medical opinion a more regular part of rural people's existence. The men who served in the army were subject to army medicine; the men conscripted after the introduction of the universal conscription law of 1872 also had regular encounters with physicians. In addition, a terrible smallpox epidemic in 1870 and 1871 helped convince the residents of the Seine-et-Oise of the usefulness of vaccination.

None of these factors, of course, would have been sufficient had not the medical profession proved that it had at least some useful knowledge and some worthwhile approaches at its disposal. And, coincidentally, the early 1880s saw the beginnings of the age of bacteriology; indeed; within the five years between 1880 and 1884, scientists had correctly identified the causative organisms of typhoid fever, cholera, tuberculosis, and diphtheria.[29] The physicians and other members of the Departmental Hygiene Council closely followed this work, and their writings on disease show their awareness of the rapid progress of medical science as they strove to interpret their data in a new light. The reception the council's work received from town councils and local people alike was more wholeheartedly positive after the spectacular successes of Koch and Pasteur. As Richard Shryock has noted, the prestige of medical science increased suddenly and dramatically in the 1880s,[30] and the Departmental Hygiene Council reaped the benefit of this gain.

The annual epidemic reports after 1880 described the same diseases, but with a fresh understanding. Typhoid fever, for example, was no longer seen as vaguely connected with dirty water, but more specifically with contaminated drinking water. Dr. Paris, in the annual epidemic report of 1882, pointed clearly to poor drinking water as the reason for typhoid in Versailles:

> As for drinking water, all hygienists agree that it is important in the propagation of typhoid fever. It is the most frequent vehicle of contagion. With a water supply fed by Paris sewer water, Versailles is in the worst possible position.[31]

Perhaps the most telling break with the earlier view of typhoid as resulting from fumes and emanations occurred in an interchange between M. Vian, the mayor of St. Chéron, and the council during its September 1892 meeting. M. Vian, who was also a deputy on the national level and was most inclined to cooperate with the Departmental Hygiene Council, volunteered to postpone the annual dredging of the Orge River, if the council felt that such action would help in the management of the typhoid epidemic. (According to the previously held miasmatic theory of disease, the dredging of a river would release noxious fumes into the air that could cause typhoid.) The council advised him to proceed with the dredging, and in a key pronouncement, observed that

> as for the epidemic of typhoid fever, its cause is completely unrelated to the [dredging of the] Orge River, and the epidemic will not end until the steps repeatedly advised by the hygiene council as a result of the reports of special commissions are implemented.[32]

These measures included the creation of several public fountains free from all contamination; this could be accomplished by tapping some springs upstream from the town. Dr. Bouillon-Lagrange, the third generation of physicians in his family to practice near St. Chéron, advised "teaching good hygiene habits (personal hygiene and domestic hygiene) and for this put up posters with guidelines or even have a public lecture on the rules of hygiene."[33] He assured the Departmental Hygiene Council that the St. Chéron town council would cooperate because, as his colleague Dr. Diard was to note two years later when typhoid recurred, "the municipality, concerned about the health of the townspeople, has put into practice all our suggestions."[34]

The attitude of local people toward medical care also appears to have changed. When typhoid fever broke out in Montlignon in the arrondissement of Pontoise in 1894, the prefect advised dispatching public health service doctors to reassure the population; during the 1820s and 1830s, his predecessors had often deferred sending such doctors for fear of spreading panic. And when Dr. Chantemesse, the assistant inspector-general of sanitary services, visited Montlignon, he noted that "the dispatch of a steam disinfecting apparatus is demanded by the townspeople, and would be welcomed with gratitude."[35]

Of the airborne diseases, diphtheria held a prominent position in the annual epidemic reports after 1880. The epidemic report for 1881 noted a considerable rise in its incidence; in 1882, it was even more widespread. With the isolation of the diphtheria bacillus in 1894, research began in earnest on methods of controlling the disease. The early 1890s saw work that suggested the possibility of serum therapy for diphtheria, and in September 1894, Emile Roux presented a paper on this subject to the Eighth International Congress of Hygiene and Demography at Bucharest. It was from this point that diphtheria antitoxin began to be employed generally.[36]

The favored position of the Seine-et-Oise, close to Paris and the vacation home for many Parisian doctors, is evident when we trace the history of the introduction of serum therapy for diphtheria into the department. In July 1894, diphtheria broke out in Lainville, a small town eight miles away from Mantes. In October, the schoolteacher's son contracted the disease, and died during an emergency tracheotomy. When the schoolteacher's second child also caught diphtheria, the distraught father appealed to a doctor who spent summers and weekends in Lainville. The doctor was able to procure the serum, and the child recovered. Word of this cure traveled fast; in the nearby town of Sailly, the father of a young diphtheria patient, after realizing that the local doctor did not have the serum, went to Paris to get it. Parents vied with each other to obtain this new miracle drug of the 1890s; Dr. Bonneau, the epidemic doctor of the arrondissement of Mantes noted that when another child in Lainville got diphtheria, "his parents, fearing that they would not be able to get serum, had the singular idea of taking him to Paris," where he was admitted to the Hospital for Sick Children, treated with serum and recovered. Dr. Bonneau deplored the transporting of sick, germ-carrying children, but had nothing but praise for the behavior of the schoolteacher who "although cruelly stricken by the death of his young son, did not hesitate to visit the sick to supervise the treatment ordered by the doctors, to take temperatures, and to note the effect of the antidiphtheria serum injections."[37]

Nor was serum the only weapon against diphtheria for which local people had any enthusiasm. They also eagerly adopted disinfecting procedures. During the diphtheria epidemic in Montmorency in the winter of 1894–1895, the municipality had provided for the disinfection of house walls with an antiseptic solution, but it had made no provision for disinfecting clothes and bedding. Apparently, concerned parents felt the need to go further, for a worried letter from the Minister of the Interior to the prefect of the Seine-et-Oise asked whether it was indeed true that "one individual, desirous of having the clothes and bedding of one of his children disinfected, sent them from Montmorency to a steam disinfecting plant . . . in Paris by using the stagecoach that provides most of the transportation between Paris and Montmorency?"[38] If this were true, continued the Minister of the Interior, the town of Montmorency should provide for complete disinfection within its boundaries. Willing use of the steam disinfecting apparatus was reported in Argenteuil and Villeneuve St. Georges.[39]

Indeed, people were more than willing to employ all manner of disinfection and fumigation; they were positively eager. Richard Shryock noted dryly that in the early days of the bacteriological era

> Enthusiasm for fumigation was revived, in the belief that all sorts of bacteria could be destroyed in this manner. A period of "promiscuous disinfection and fumigation" ensued, which involved more than sailors and shipping. School children were liable, on any morning, to find their eyes smarting and the classroom air heavy with fumes of one sort or another.[40]

A similar example of popular enthusiasm for medical treatment occurred in Flins in 1881. Sweating sickness had broken out there, and public health authorities initially dispensed smallpox vaccination, even though this was not the ideal treatment for the disease (none was known). Dr. Paris, in response to popular pressure, advised that the vaccination program be continued, "since it can raise the morale of the people who are very stirred up."[41] Apparently medical intervention, any medical intervention, had a positive effect.

To a certain extent, the people of Flins had recourse to medication in much the same way their ancestors may have had recourse to magic. But there are examples of calmer, more rational action as well. As typhoid raged through Louveciennes in 1892, the town council, responding to public pressure, urged a formal inquest into the cause of the disease: no alarm here, simply a desire for more information.[42] Two years later, when typhoid broke out in Montlignon, the townspeople there, too, called for more medical help.[43] And we have already seen the enthusiastic response afforded diphtheria serum.

It is important not to overstate the case. Although many times medical help was sought out, often it was not. Dr. Fritz and Dr. Prestat of the arrondissement of Corbeil did succeed in performing many revaccinations in 1881, but they "found resistance in the ignorance of the peasants who preferred preserving powders given out by a neighboring pharmacist."[44] Still, even a mixture was progress. A survey of 1957, conducted by IFOP, the French national opinion poll, on "The Frenchman and His Doctor," revealed a mixture of attitudes, of trust and of suspicion, that resemble the feelings of the people of the Seine-et-Oise around 1900.[45] It is this ambivalence, containing at least partial trust, that brings the people of the Seine-et-Oise in the last decade before World War I as close to us as to their forebears of Napoleon's time.

One sort of medical facility that received only grudging popular acceptance before World War I was the hospital. To our minds, the hospital represents a totally medical environment, a useful setting for specialized medical care. During the nineteenth century, physicians and administrators often expressed similar views. For them, a hospital setting had the advantage of isolating the patient from what was considered the deleterious, distracting influence of family and friends. The mayor of Argenteuil, for example, was simply reflecting current medical opinion when, in 1832, he pointed out the superiority of hospital care for cholera over home treatment:

> Once the initial panic had worn off . . . , relatives and neighbors flocked to the bedsides of the sick who were at home, but instead of hiding their fears, they moaned and wailed out loud and in their heedlessness openly predicted the death of their friends. The cholera patient conserves his reason until the very end, and hearing the things people said around him, he exaggerated the gravity of his condition and lost all hope. . . .
>
> In the hospital, on the contrary, from which I had barred even the closest relatives and where there were only two doctors, the nursing sisters, and the other staff, there reigned the greatest calm. Far from stirring up the imaginations of the sick, we were only full of consolation, and we removed any discouraging influences.[46]

Dr. Bonneau held similar beliefs about the advantages of hospital care at Mantes; in his annual report for 1861, he noted that "country people themselves are beginning to understand that long and serious diseases have a greater chance of being cured in an environment where they are surrounded with intelligent and continual care."[47] Much like the mayor of Argenteuil, Dr. Bonneau looked to the hospital to create a total environment more sympathetic to medical care than a rural village.

But for rural people of the nineteenth century, to go to the hospital was to admit defeat. The hospital was a place reserved for those without family, friends, or hope. When Edmonde Charles-Roux, in her appealing biography of the designer Coco Chanel, wanted to give the reader a sense of the deep-rooted despair and alienation of the Chanel family in the nineteenth century, she could think of no better item from their family history than the fact that for several generations, successive Chanels were born at the hospital.[48]

This attitude persisted with remarkable tenacity until World War I, as we can see by a computer analysis of the patient population of one typical hospital in the Seine-et-Oise, the hospital of Mantes. Information on about 9000 patients who entered the hospital between 1800 and 1911 was coded onto IBM cards, and a computer analysis was performed. The patient registers from which the data were taken contain considerable information about each person admitted: his name, marital status, age, birthplace, residence, occupation, length of stay, and whether death occurred. For the registers after 1895, there is information about the diagnosis and about who paid for the hospitalization.

Now, it can easily be argued, computer printouts are a cold and heartless way of assessing people's feelings about pain, suffering, and sometimes death. But since these patients were inarticulate—they did not leave diaries, letters, or other records of their feelings for the historian to scrutinize, we are forced to try to infer from their actual behavior—in this case the ways they made use of the hospital—how they may have felt about hospital care.

During the first quarter of the nineteenth century, the number of beds in the hospital tripled, rising from eighteen to fifty-four. Between 1800 and 1825, from thirty-three to ninety-nine patients entered the hospital each year. They generally spent from two weeks to two months there. Men outnumbered women by three to two. Most of the patients recovered, and relatively few of them made repeat visits to the hospital.

The picture of patient behavior in 1839–1853 is very different. Men outnumbered women by a much higher proportion, four to one. Patients were somewhat younger than they had been in 1800–1825. Most striking was the enormous rise in the number of admissions per year; 48 had been the average for 1800–1815 and 70 for 1816–1825. But for 1839–1845, the average number of admissions per year was 130. In 1846, there were 287 entries, in 1847, 507, in 1848, 437.

The hospital was able to accommodate all these people not by doubling them up in beds (throughout the period one patient to a bed had been the rule), but because the length of time patients stayed in the hospital dropped precipitously. Between 1847 and 1853, the years when the numbers of admissions were at their highest, between half and two-thirds of all patients stayed less than a week. Many of them simply stayed overnight.

Miraculous cures at Mantes? Hardly. Another salient feature of these short-term patients is that the great majority of them were travelers, people without a fixed domicile. The hospital, by its own rules and ordinances, was required to care for them. As bread prices rose in the late 1840s, so too did the number of these travelers who came to the hospital, sick or perhaps simply hungry. In 1840, only twenty-one patients, 14 percent of that year's total, were travelers; in 1847, 327, or 64 percent of the people admitted, fell into that category. And these travelers were not simply the product of what demographers call micromobility; some of them had come a long way indeed: book peddlers from southern France, picture merchants from Alsace. Most patients, even most travelers, were day laborers or artisans in fairly standard trades, but the frequency of people skilled in very particular crafts among the travelers is striking. These travelers almost never died in the hospital; they had a mortality rate of less than 2 percent.

The vagabonds must have imparted a certain color to the hospital. Whether they made it "the liveliest spot in town," the phrase Ernest Renan used to describe the hospital of his home town in Brittany, is another question.[49] What is clear is that the hospital administrators had little love for these homeless people. In the general report of 1860, the administrators noted with relief that the hospital population had declined in recent years, largely because of steps taken "to admit only people who really needed treatment and to turn away a considerable number of travelers who, instead of making themselves useful, are often disposed to abuse help and whose constant presence in the wards bothered the sick who really need care."[50] In other words, the unworthy poor were to be turned away. The hospital administrators were much more sympathetic when the number of patients rose again in 1861 because of "the lack of work and high food prices."[51] The hospital, then, was still seen as a charitable institution, to help the worthy sick poor.

Another group whose use of the hospital was followed with great interest by its administrators were the peasants of the arrondissement of Mantes. A special legacy of 1848 had endowed five beds for their exclusive use. In 1857, after the beds had been in use for six years, the administrators worked out some statistics about their use. One hundred forty-one people from forty-five villages had taken advantage of them. Mortality among the rural people was about 15 percent, or twice that of the rest of the hospital population during the same years. Even Dr. Bonneau, in the middle of an extremely sanguine report on the hospital's medical service in 1861, could not avoid the significance of this higher death rate. He wrote:

> We must recognize that most of these sick peasants come to our establishment only to die there, almost always afflicted with chronic, incurable diseases like cancer, heart trouble, paralysis, tuberculosis; they come only after having exhausted their pecuniary resources and other possibilities of cure. They take advantage of the . . . legacy only to try out new remedies advised by a new doctor.[52]

By the turn of the twentieth century, the Mantes hospital had grown and developed. The level of sophistication of care it provided apparently improved. The seventy-one beds of 1901 were divided among four separate services. Thirty-nine beds were on the medical service, eighteen on the surgical service, twelve in a special isolation unit for contagious disease cases, and two for psychiatric cases. The contagious disease unit was a development of the post-1850 period, as was the psychiatric division. Back in the 1830s, the hospital had constructed two padded cells for insane patients; by the turn of the century, however, psychiatric cases were kept only for observation, and then sent on to the larger hospital for the insane at Clermont. In 1891, the operating room was completely renovated, and in the early years of the twentieth century, the hospital hired a doctor to do anesthesiology to round out its staff of two other physicians.

Yet despite these improvements, the hospital still attracted only the poorer, more rootless people of the Mantes area. Only 35 percent of the adults who entered the hospital between 1895 and 1911 were married; this figure does not correspond to the percentage of unmarried adults in the population at large. Surprisingly few patients had been born in the region: only 8 percent in Mantes itself, and 23 percent in other parts of its arrondissement. Fewer than a third, in other words, had local roots older than one generation. By contrast, almost 15 percent of the patients had been born in one of the five departments that formed the old province of Brittany, a backward, impoverished area that saw great emigration to other parts of France in the late nineteenth and twentieth centuries. By the criteria of both marital status and birthplace, then, the hospital patients appear to be a more isolated group than the residents of the area as a whole.

People from different occupational groups came to the hospital with somewhat different ailments.[53] Between 1895 and 1911, about 2 percent of the patient population could be viewed as middle class: that is to say, bureaucrats, professionals, holders of municipal jobs, wealthy or retired people, and highly skilled workers. Although this small cohort's diagnoses occurred in roughly the same proportions as those of the patient population at large, there were several significant differences. The frequency of neurological, psychiatric, and alcohol-related illnesses was almost three times higher among middle-class patients than in the overall patient population. Middle-class families may have protected their members from the indignities of hospital care for many diseases, for childbirth, even for surgery, which was often performed at home. But once a middle-class individual exhibited behavior that was considered deviant, or simply unpleasant to look at, as in the case of some neurological disorders, the possibility of hospitalization increased. By contrast, extremely few agricultural workers were hospitalized for alcohol-related, psychiatric, or neurological reasons; unless a person was really out of control and a danger to others, rural villagers tended to be extremely tolerant of personality aberrations.

But it was not only the families of the comparatively wealthy or well-trained that regarded them in a different light; so did the doctors at the hospital. Several of the diagnoses in the hospital registers were fairly sophisticated statements about liver and gall bladder dysfunction, or about metabolic and endocrine problems. These diagnoses almost invariably were those of middle-class patients. Drs. Bonneau and Bihorel, perhaps, when they saw a hobo enter the hospital, took him at his word that he was "under the weather," "all bent over." They saw his poverty first, and did not always look behind it to search for the more precise form of the illness. The diagnosis for 141 people was simply the peasant word for aches and pains, *courbature,* and it is striking that two-thirds of the people who received this diagnosis were vagrants. A more highly placed patient, who possibly was paying at a high rate for his care, often received a more detailed assessment.

Indeed, one of the most salient characteristics of the medicine at the Mantes hospital is precisely this wide range of conditions treated and the great variation—from crude to quite sophisticated—in the quality of the diagnosis. If maternity patients are excluded, around a quarter of all patients were admitted because of what can really be called the occupational hazards of being a marginal rural worker. Trauma, orthopedic injuries, "aches and pains," problems resulting from neglect and exposure, chemical poisoning— these accounted for half the hospitalizations of all the vagrants studied, but many Mantes residents and rural people came in for these reasons, too. On the other hand, more refined diagnoses were made in the areas of urology, cardiology, oncology, neurology, and psychiatry. The doctor in 1910 who diagnosed osteosarcoma of the right knee, for example, had looked at his patient very closely, and with a trained eye; the doctor who diagnosed conversion hysteria of the leg during the same year surely knew the work of Charcot, not to mention Freud.

Despite its well trained staff, the Mantes hospital, like its counterparts all over France, was underused on the eve of World War I. The traditional association in the popular mind between the medical care of the hospital and the custodial charity of the poorhouse was still too powerful to enable people to make rational use of hospitals. This should serve to remind us that problems of the dispensing of medical care are as much problems of mentality as of technology. It was only after 1880 when the world views of rural people and physicians began to converge, however slightly, that medical care was accorded some degree of acceptance. And even today, the media abounds in articles and discussions that remind us of the continuing need to adapt medical technology to our own emotional, cultural, and ethical needs.

NOTES AND REFERENCES

1. "Traitement des épidémies," in *Receuil des circulaires et instructions émanées du ministère de l'intérieur de 1790 à 1830 inclusivement* (Paris, 1850): I, 219.
2. *See* HANNAWAY, C. C., "The Société royale de médecine and Epidemics in the Ancien Régime," *Bulletin of the History of Medicine* **46:** 257–273, 1972, for background on the handling of epidemics in the eighteenth century.
3. FRANK'S *System einer vollständigen medicinischen polizey* was published between 1779 and 1819.
4. The grade of health officer, a subordinate sort of doctor authorized to practice in one department only, was created by the law of 19 ventôse *an* XI (March 10, 1803) that regulated medical practice in France for most of the nineteenth century.
5. "Traitement des épidémies," in *Receuil des circulaires . . .* , I: 221.
6. Archives Départementales des Yvelines (hereafter abbreviated as ADY), 9 M 204.
7. ADY, 7 M 22.
8. These epidemic doctors were not salaried officials; rather, they were reimbursed for each trip to an affected area. In this way they differ from the regularly salaried cantonal physicians in the department of the Bas-Rhin at the same period. *See* SUSSMAN, G. D. "Enlightened Health Reform, Professional Medicine and Traditional Society: The Cantonal Physicians of the Bas-Rhin, 1810–1870," *Bulletin of the History of Medicine* **51:** 565–584, 1977.
9. ADY, 7 M 56, intermittent fever at Prunay, 1802.
10. ADY, 7 M 56, various contagious diseases and epidemics at Moigny.

11. ADY, 7 M 56, measles at Ville d'Avray, 1842.
12. ADY, 7 M 56, report of Dr. Lelarge on the epidemic at Gagny, 1807.
13. ADY, 7 M 56, intermittent fever at Gironville, 1839.
14. ADY, 7 M 56, epidemic in the cantons of Isle Adam and Luzarches, 1821.
15. ADY, 7 M 56, fevers at Gif, 1811.
16. ROSENBERG, C. E., "The Therapeutic Revolution: Medicine, Meaning, and Social Change in Nineteenth Century America," *Perspectives in Biology and Medicine* **20:** 485–507, 1977.
17. ADY, 7 M 50, letter from Dr. Fossoyeux to the prefect, August 10, 1832.
18. The relatively high mortality from cholera stood in marked contrast to the much lower incidence of death from other diseases in the Seine-et-Oise. *See* DOWLING, H. E., *Fighting Infection: Conquests of the Twentieth Century,* Cambridge, MA and London, Harvard University Press, 1977, p. 12.
19. ADY, 7 M 50, Rapport adressé à M. le préfet de Seine-et-Oise sur le choléra-morbus épidémique qui a régné dans l'arrondissement de Mantes pendant les mois d'avril, mai, juin, juillet, août et septembre 1832 par M. le docteur Dussaux.
20. ADY, 7 M 50, Rapport fait et adressé par Louis-Jérôme Fossoyeux, ancien médecin des armées attaché au service des hôpitaux et hospices civils de Paris, résidant à Sarcelles, nommé officier de santé des épidémies du canton d'Ecouen, à M. le Conseiller d'état préfet du département de Seine-et-Oise, October 10, 1832.
21. ADY, 7 M 50, Rapport de Monsieur Edouard Petit, Docteur en Médecine de la Faculté de Paris, Chevalier de la Légion d'Honneur, Médecin des Epidémies pour la Sous-Préfecture de Corbeil, Médecin et Administrateur honoraire de l'Hospice Civil, Associé correspondant de l'Académie Royale de Médecine, de l'Académie de Dijon, de la Société Médicale d'Emulation de Paris, etc., etc., à Monsieur le Préfet du département de Seine-et-Oise, May 5, 1834.
22. ADY, 7 M 50, Rapport adressé à M. le Préfet de Seine-et-Oise sur le choléra-morbus épidémique . . . par M. le docteur Dussaux. Similar hostility was observed throughout France; *see,* for example, LEONARD, J., *La vie quotidienne du médecin de province,* Paris, Hachette, 1978, p. 260.
23. ADY, 7 M 50, letter from Dr. Peyron to the prefect, July 25, 1832.
24. ADY, 7 M 53, Conseil d'hygiène et de salubrité: Enquête sur la marche et les effets du choléra, arrondissement de Corbeil. Jacques Léonard found similar behavior during the 1865 cholera epidemic; *see* LÉONARD,[22] p. 260.
25. *L'Union médicale,* September 17, 1850.
26. ADY, 7 M 2, Departmental Hygiene Council (hereafter abbreviated as DHC) meeting of March 17, 1852.
27. ADY, K, March 25, 1850.
28. These figures are from the commune of Bonnières-sur-Seine.
29. In 1880, Eberth discovered the typhoid bacillus; his findings were confirmed by Koch in 1881 and by Gaffsky in 1884. In 1882, Koch identified the tuberculosis bacillus, and two years later, he isolated the cholera bacillus. In 1884, Loeffler cultured the diphtheria bacillus.
30. SHRYOCK, R., *The Development of Modern Medicine: An Interpretation of the Social and Scientific Factors Involved,* New York, Knopf, 1947, pp. 336–355.
31. ADY, 7 M 58, Rapport sur les maladies épidémiques et contagieuses qui ont régné dans le département de Seine-et-Oise pendant l'année 1882 par le Docteur Paris, médecin des épidémies de l'arrondissement de Versailles.
32. ADY, 7 M 57, DHC meeting of September 28, 1892.

33. ADY, 7 M 57, letter from Dr. Diard to the subprefect of Rambouillet, August 23, 1890.
34. ADY, 7 M 57, DHC meeting of September 29, 1892.
35. ADY, 7 M 57, Epidémie de fièvre typhoïde à Montlignon. Rapport de M. le Docteur Chantemesse, inspecteur-général adjoint des services sanitaires, August 6, 1894.
36. ROSEN, G., "Acute Communicable Diseases," in *The History and Conquest of Common Diseases,* Walter R. Bett, Ed. Norman, OK, University of Oklahoma Press, 1954, p. 19.
37. ADY, 7 M 57, report of Dr. Bonneau on diphtheria at Lainville, 1894.
38. ADY, 7 M 57, letter from the Minister of the Interior to the prefect of the Seine-et-Oise, December 1, 1894.
39. ADY, 7 M 57, letter from the mayor of Argenteuil to the prefect of the Seine-et-Oise, April 22, 1895; letter from the mayor of Villeneuve-St.-Georges to the subprefect of Corbeil, December 19, 1893.
40. SHRYOCK,[30] p. 291.
41. ADY, 7 M 8, DHC meeting of February 23, 1881.
42. ADY, 7 M 57, letter of December 20, 1892 from the mayor of Louveciennes to the prefect.
43. ADY, 7 M 57, letter of August 2, 1894 from the prefect to the Minister of the Interior.
44. ADY, 7 M 8, Rapport sur les maladies épidémiques et contagieuses qui ont régné dans le département de Seine-et-Oise pendant l'année 1881 par le Docteur Paris, médecin des épidémies de l'arrondissement de Versailles.
45. BOLTANSKI, L., *Prime éducation et morale de classe,* Paris, Mouton, 1969, p. 97.
46. ADY, 7 M 50, Quelques réflexions sur le choléra-morbus soumises à M. le préfet de Seine-et-Oise par le maire d'Argenteuil, November 22, 1832.
47. Archives de l'hôpital de Mantes (hereafter abbreviated as AHM), Rapport sur le service médical de l'hôpital de Mantes pendant l'année 1861.
48. CHARLES-ROUX, E., *L'irrégulière ou mon itinéraire Chanel,* Paris, Grasset et Fasquelle, 1974.
49. RENAN, E., *Souvenirs d'enfance et de jeunesse,* Paris, 1876, p. 18.
50. AHM, Compte moral administratif of 1860. Observations générales.
51. AHM, Compte moral administratif of 1861. Observations générales.
52. AHM, Rapport sur le service médical de l'hôpital de Mantes pendant l'année 1861.
53. A disease classification chart, applicable to these turn-of-the-century patients, was kindly prepared by Enid A. Lang, MD.

The Developing Technology of Apparatus in Psychology's Early Laboratories[a,b]

FAIRFID M. CAUDLE

Department of Psychology, Sociology and Anthropology
The College of Staten Island
The City University of New York
Staten Island, New York 10301

During the centennial year of the founding in 1879 of the first laboratory devoted to experimental psychology, psychologists have been celebrating the remarkable growth of their discipline. The term "psychology" now includes areas as diverse as the biochemical basis of behavior and industrial management, as broad as child development, and as circumscribed as experimental hypnosis. Today, the experimental psychology of the laboratory is but one small portion of a very large pie. However, it was not always thus—a century ago, experimental psychology *was* psychology, and, in establishing a scientific psychology, an indispensable role was played by the instruments employed in early laboratory research. Rather than focusing upon the major personages, systems, and theories of the first century of experimental psychology, it is instead the instruments themselves that I wish to consider in this discussion.

In 1867, William James, then in Berlin, noted the following in a letter to Thomas W. Ward:

> It seems to me that perhaps the time has come for psychology to begin to be a science—some measurements have already been made in the region lying between the physical changes in the nerves and the appearance of consciousness . . . and more may come of it. I am going on to study what is already known, and perhaps may be able to do some work at it. Helmholtz and a man named Wundt at Heidelberg are working on it, and I hope I live through this winter to go to them in the summer. (W. James, 1867/1920)

Some twelve years later, in 1879, Wilhelm Wundt (see FIG. 1) established the first psychology laboratory in Leipzig, Germany. Wundt and those who studied with him were eager to establish a science of psychology, and one route to this goal was to emulate at least in part the methodologies of other, well-established sciences. In particular, an important goal of early experimenters in psychology was to achieve precision in quantification and measurement

[a]This paper was presented at the October 22, 1980 meeting of the Section of History, Philosophy and Ethical Issues of Science and Technology of the New York Academy of Sciences. An earlier version was presented as a Centennial Lecture with the title "Laboratory Instruments and History: Comforting Continuities (and Doleful Dead Ends)" at the Annual Convention of the American Psychological Association in Montreal, Canada, in September, 1980.

[b]The research on which this paper is based was supported in part by research awards from The College of Staten Island of The City University of New York, and the Archives of the History of American Psychology at The University of Akron in Akron, Ohio.

0077–8923/83/0412–0019 $01.75/2 © 1983, NYAS

FIGURE 1. Wilhelm Wundt in a popular photograph thought to have been made about 1870 (Bringmann, Ungerer & Ganzer, 1980).

comparable to that which had thus far been gained in the physical and biological sciences.

Thus, early psychologists enthusiastically borrowed and adapted the scientific instruments that had heretofore been used to explore problems related to the physical laws of acoustics and optics and the physiology of the sense organs (Davis and Merzbach, 1975). However, there was an important difference: such instruments now began to be used in investigating *psychological* problems such as the contents of consciousness, the processes of perception, and the time required for mental processes to occur. In these early efforts, psychologists not only adapted existing instruments; they also designed their own and the era of "Brass Instrument Psychology" was born, offspring of a marriage of sorts between some of the questions and issues raised long before by philosophers and the empirical methodology painstakingly devised by those within the physical and biological sciences.

Imagine for a moment that it could be possible, through some quirk of time, for Wundt to visit a modern psychology laboratory. We might expect it to be unrecognizable to him, at least at first, for he would be confronted with a gleaming array of sophisticated devices such as electronic relay boards, biofeedback apparatus, and microprocessors. Superficially, at least, there would appear to be little in common between the apparatus employed during the so-called era of brass instruments and the electronic era of today.

To be sure, the early laboratories contained instruments no longer to be found and today's laboratories contain some devices that our founding fathers (and mothers) would have gasped at in amazement. However, once Wundt, during this hypothetical visit, began to look beyond the cosmetic aspects of the instruments in order to concentrate instead upon their *functions* and *applications,* I suspect that he would find a surprising degree of continuity between the objectives and methods developed earlier in the first century of experimental psychology and those in use today.

THE MEASUREMENT OF REACTION TIME

For example, consider the problem of precision in timing. The development and adaptation of apparatus to measure small intervals of time was a central goal in early psychology laboratories, and users of instruments of all types have always had to contend with error in measurement, whether in astronomy, physiology, or psychology. I am sure that most of you are familiar with how one impetus to the investigation of reaction time as a psychological problem grew out of the discovery of the "personal equation," the circumstance that, because of individual differences, different persons observing stars would get different results in measuring the time at which a particular star made its transit across parallel lines in a telescope. This source of error in observation led to the realization that mental processes are not instantaneous and eventually gave rise to the psychological study of reaction time (Boring, 1950).

The first real study of reaction time was the complication experiment. Wundt borrowed the term "complication" from Herbart who used this term to mean a mental operation that involved processes from more than one of the senses (Boring, 1950). Wundt initially believed that he had invented a seemingly foolproof way to measure the time required for mental processes to occur. Referring to his complication apparatus (FIG. 2) and the personal differences that had been observed by astronomers, Wundt noted that these

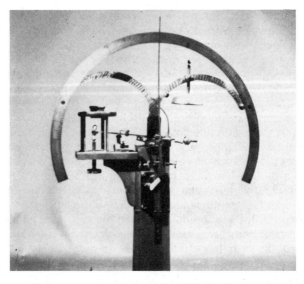

FIGURE 2. Complication apparatus, designed by Wilhelm Wundt (photographed by the author).[c]

"can only be explained by the fact that the process of perception and thinking requires a certain amount of time" (Wundt, 1862, quoted by Diamond, 1974, p. 697). Wundt described the operation of the complication apparatus as follows:

> Recently I have tried to determine this time with greater accuracy, experimentally. I let a pendulum swing along a circular scale; at a certain point in its path it strikes a [hidden] lever. We can then compare the true position on the scale at the moment when the sound was produced with the position where it seemed to be when the sound was heard. A constant scale difference was found, and from this and the speed of the pendulum the time elapsed between the auditory and visual perceptions was calculated. This time is found to be, on the average, 1/8 second. . . . I believe that this investigation of the speed of perception can be called *purely* psychological. (Wundt, 1862, quoted by Diamond, 1974, p. 697).

With this apparatus, Wundt initially felt that he had discovered a "psychic constant" (p. 697) which he believed to be the "swiftest thought" (p. 698). Diamond (1974) has noted that the claims Wundt made for the significance of the complication experiment were not repeated in future writings. Nevertheless, the measurement of the speed of mental processes became entrenched in laboratory psychology, although with, perhaps, more realistic expectations of what could be achieved.

---

[c]All illustrations with the notation "photographed by the author" are of instruments in the collection of the Archives of the History of American Psychology, The University of Akron, Akron, Ohio.

The complication apparatus was only one of many devices employed in studies of reaction time. An essential device in many studies was the Hipp chronoscope, an electromagnetic device used to measure small intervals of time (FIG. 3). Numerous types of chronoscopes were invented; some had electromagnetic mechanisms; others were regulated by the swing of a pendulum; and at least one was activated by an electromagnetic tuning fork. Each of them had its own particular applications but all were designed to enable precision in the measurement of small time intervals.

Reaction time experiments presented other problems in addition to accuracy. Not only had elapsed time to be measured accurately, but a means of recording what transpired *during* time was also needed. In order to obtain written records, graphic recording devices were borrowed and adapted from the physiologists and the kymograph came into its own.

The kymograph was an instrument that recorded the course and duration of any physiological or muscular process. It consisted of a drum revolving at a uniform speed, which was covered with a record sheet (usually of smoked paper) on which a stylus made a written record of the process being measured (FIG. 4).

The tuning fork was often used in conjunction with a kymograph, because of the precision with which a tuning fork oscillated and the fact that it could be

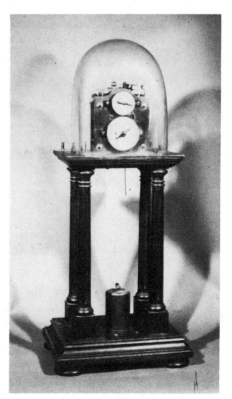

FIGURE 3. Hipp chronoscope (photographed by the author).

FIGURE 4. One version of the kymograph (photographed by the author).

FIGURE 5. Apparatus employing a tuning fork to divide a second into hundredths. The fork is run by the lamp battery at the left, while the lamp battery at the right activates the magnetic coil to which it is connected (Scripture, 1897).

used to divide a second into precise parts. One arrangement using the tuning fork to divide a second into hundredths is shown in FIGURE 5. In a separate account, Scripture has described this general process as follows:

The first thing to be done is to set up a tuning-fork—not a little one, such as musicians carry in the pocket, but one a foot long, vibrating one hundred times a second. By means of a battery and a magnet this fork is kept going of itself as long as we please. The prongs of the fork move up and down one hundred times a second. Every time the lower prong moves downward, a point on the end dips into a cup of mercury, whereby an electric circuit is closed. This electric circuit passes through a little instrument called a time-marker, which makes a light

FIGURE 6. An early voice key (photographed by the author).

pointer move back and forth also one hundred times a second. The point of the time-marker rests on a surface of smoked paper on a cylindrical drum. . . . To preserve the record, *i.e.,* to keep the smoke from rubbing off, the paper is cut from the drum, run through a varnish, and dried, the result being what might well be called a study in black and white. (Scripture, 1895, pp. 29–30)

Another form of apparatus employed in reaction time experiments was the voice key (FIG. 6), in which sound vibrations caused a membrane to vibrate and this in turn caused an electrical circuit to close and activate a timing device. Some voice keys used technology developed from the telephone.

These were but a few of the tools employed in the measurement of reaction time and the duration of various processes. In the modern laboratory, precision in timing continues to be a major goal, and there is as large a variety of timers today as in earlier laboratories. These timers (e.g., FIG. 7) are not as elegant as those of our predecessors—brass and polished wood have regrettably given way to plastic and stainless steel—but their functions remain much the same.

Today we have highly accurate methods of determining reaction time. The ugly duckling of the early voice key has developed into a sleek swan, the sensitive and much-streamlined electronic voice reaction time apparatus (FIG. 8).

FIGURE 7. A modern interval timer (reproduced with the permission of the Lafayette Instrument Company, Inc.).

Through the years the kymograph has been greatly modernized. I am sure that those who laundered the clothing of early psychologists were grateful when the smoked paper used for kymograph drums was replaced by other surfaces such as waxed paper and then was replaced entirely in sophisticated recorders which use the less cumbersome and more reliable pen and paper (FIG. 9).

However, what has really changed? While today's psychologists are not so blissfully convinced that the exact time of an idea can be measured (although they have by no means given up trying), and although today's tools are more precise and reliable, their functions remain much as they were in the early

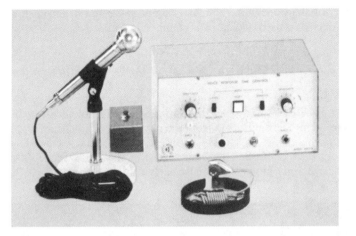

FIGURE 8. Electronic voice reaction time apparatus (reproduced with the permission of the Lafayette Instrument Company, Inc.).

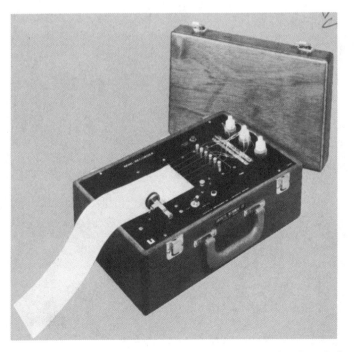

FIGURE 9. A six-channel event recorder, a modern instrument with functions similar to the kymograph (reproduced with the permission of the Lafayette Instrument Company, Inc.).

days of psychology. To be sure, we have found some new problems to study but much in our basic methodology has changed but little.

APPARATUS FOR AUDITORY STUDIES

What of other problems explored in the laboratories of our predecessors? Since many of the early laboratory studies of psychology grew out of the physics of optics and sound and from the physiology of the sense organs, it was natural that major efforts in early studies were directed toward understanding the senses, with vision and hearing receiving the lion's share of attention. What role did apparatus play in these early studies?

FIGURE 10. Hermann L. F. Helmholtz (reproduced from Helmholtz, 1863/1954).

FIGURE 11. Part of a series of tuning forks, mounted on resonance boxes (photographed by the author).

The major impetus to early auditory studies came, of course, from Helmholtz (FIG. 10). In the preface to his 1863 work "On the Sensations of Tone" Helmholtz wrote:

> In the present work an attempt will be made to connect the boundaries of two sciences, which, although drawn towards each other by many natural affinities, have hitherto remained practically distinct—I mean the boundaries of *physical and physiological acoustics* on the one side, and of *musical science and esthetics* on the other. (Helmholtz, 1863/1954, p. 1)

Helmholtz sought to explore relations between the physical stimuli, tones, and the psychological sensation and perception of them. Helmholtz worked closely with the instrument maker Koenig to design beautifully crafted and precise tools with which to investigate the problems that he formulated. Among the instruments adapted by Helmholtz was, of course, the tuning fork, used to produce a tone of a desired pitch (FIG. 11). Tuning forks appeared in many sizes, sometimes mounted on resonance boxes or being placed adjacent to resonance cylinders for greater volume (FIG. 12).

Perhaps the most beautiful of the instruments designed by Helmholtz (but, sadly, also one of the first instrumental "dinosaurs" to disappear from the active research scene) were his resonators (FIG. 13). These were devices for magnifying the intensity of a tone of some given pitch and were used to isolate simple sounds and establish the frequencies of musical tones. Concerning the goals Helmholtz had set for himself, Helmholtz wrote:

> As our problem is to study the laws of the sensation of hearing, our first business will be to examine how many kinds of sensation the ear can generate, and what

FIGURE 12. Tuning fork with resonance cylinder and mallet.

differences in the external means of excitement or sound, correspond to these differences of sensation. (Helmholtz, 1863/1954, p. 7)

Helmholtz described the resonator as follows:

These are hollow spheres of glass or metal ... with two openings ... One opening (a) has sharp edges, the other (b) is funnel-shaped, and adapted for insertion into the ear. This smaller end I usually coat with melted sealing wax, and when the wax has cooled down enough not to hurt the finger on being touched, but is still soft, I press the opening into the entrance of my ear. The sealing wax thus moulds itself to the shape of the inner surface of this opening, and when I subsequently use the resonator, it fits easily and is air-tight. (p. 43)

As Helmholtz went on to describe the operation of the resonator,

The mass of air in a resonator, together with that in the aural passage, and with the tympanic membrane ... itself, forms an elastic system which is capable of vibrating in a peculiar manner ... If we stop one ear (which is best done by a plug of sealing wax moulded into the form of the entrance of the ear), and apply a resonator to the other, most of the tones produced in the surrounding air will be

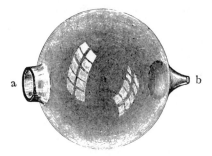

FIGURE 13. Drawing of a Helmhotz resonator (reproduced from Helmholtz, 1863/1954).

considerably damped; but if the proper tone of the resonator is sounded, it brays into the ear most powerfully. Hence, any one, even if he has no ear for music or is quite unpractised in detecting musical sounds, is put in a condition to pick the required simple tone, even if comparatively faint, from out of a great number of others. (pp. 43–44)

Although the resonator was used to assist the ear in hearing selected sounds, other instruments were employed to produce auditory stimuli with desired characteristics. There was a wide variety in the sorts of devices used to produce tone stimuli. In addition to the ubiquitous tuning fork, devices found in early laboratories included the chord siren (FIG. 14), which had a rotating

FIGURE 14. A chord siren, used to produce tones of particular frequency (photographed by the author).

disk with perforations through which blasts of air or steam were forced from a nozzle. According to Boring (1942), "the siren had the advantage of being easy to control and calibrate, since its frequency is simply the number of holes passing the air-jet in a second" (p. 328).

One versatile tone source was the whistle developed by Sir Francis Galton for determining the upper limits of hearing. As Galton described his experiences with the whistle (FIG. 15),

I contrived a small whistle for conveniently ascertaining the upper limits of audible sound in different persons.... On testing different persons, I found there was a remarkable falling off in the power of hearing high notes as age advanced. The persons themselves were quite unconscious of their deficiency so long as their sense of hearing low notes remained unimpaired. It is an only too amusing experiment to test a party of persons of various ages, including some rather elderly and self-satisfied personages. They are indignant at being thought deficient in the power of hearing, yet the experiment quickly shows that they are absolutely deaf to shrill notes which the younger persons hear acutely, and they commonly betray much dislike to the discovery. (Galton, 1883, pp. 26–27)

Not content with exploring the upper limits of hearing in humans, Galton also conducted experiments with animals:

I have tried experiments with all kinds of animals on their powers of hearing shrill notes. I have gone through the whole of the Zoological Gardens, using an apparatus arranged for the purpose. It consists of one of my little whistles at the end of a walking-stick—that is, in reality, a long tube; it has a bit of india-rubber pipe under the handle, a sudden squeeze upon which forces a little air into the whistle and causes it to sound. I hold it as near as is safe to the ears of the animals, and when they are quite accustomed to its presence and heedless of it, I make it sound; then if they prick their ears it shows that they hear the whistle; if they do not, it is probably inaudible to them. (p. 27)

FIGURE 15. Galton whistle, used for determining the upper limits of hearing (photographed by the author).

Galton went on to note that cats, small but not large dogs, and some ponies could hear the whistle but that his experiments with insects had been failures. His whistle found its way into many of the early laboratories, where it was used in more systematic research.

The tuning fork, the siren, and the whistle, as well as other devices that cannot be discussed here because of space limitations, were useful for situations where a tone source of a particular *frequency* was needed. However, they had other difficulties associated with their use. Referring to the tuning fork, Boring (1942) made an observation that pertains to other tone sources as well:

The tuning fork was valued because it gave an approximately pure tone of accurate pitch—provided it was not struck too vigorously. The fork, however, had the disadvantage that, while accurate as to frequency, it was not easily controlled in intensity. Psychologists of the nineteenth century, however, stress-

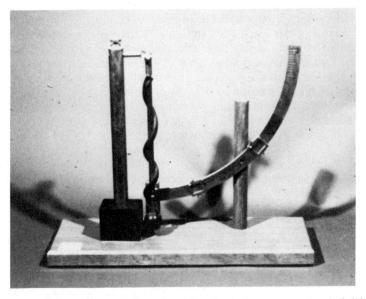

FIGURE 16. Sound pendulum, which regulated the volume of sound by varying the height from which the pendulum was dropped (photographed by the author).

ing sensory quality as more important than intensity, scarcely recognized this gross defect. (p. 330)

Although many auditory studies in early laboratories could be carried out with devices whose frequency could be controlled, psychologists also attached high priority to the measurement of auditory thresholds, or the degree of

FIGURE 17. Lehmann's acoumeter, used to measure the least perceptible intensity of sound by adjusting the height from which a pellet was allowed to fall (photographed by the author).

loudness at which it was possible to hear a stimulus. Such studies were influenced by Fechner and his description of the psychophysical methods, and, in order to determine auditory thresholds, the control of stimulus intensity was essential. Our predecessors attempted to control loudness through devices such as the sound pendulum (FIG. 16), which consisted of a hard rubber pendulum dropping against an ebony block. Loudness was regulated by the height from which the pendulum was dropped. Another widely used device was Lehmann's acoumeter (FIG. 17), used to measure the least perceptible intensity of sound by adjusting the height of fall of a pellet upon a glass surface.

One disadvantage of devices such as the sound pendulum and the acoumeter was that results varied according to the position of subjects within

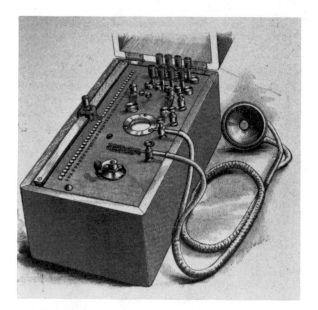

FIGURE 18. Seashore's audiometer (Whipple, 1910).

rooms and according to the acoustic qualities of the experimental rooms themselves. Seashore's audiometer (FIG. 18) was one attempt to permit testing at the ear itself in order to minimize disturbing external noises. It employed an electrically maintained tuning fork directed into a telephone receiver, with the intensity varied by means of a potentiometer (Warren, 1934).

The all-pervasive goal of precision was not really to be achieved until the development of the thermionic vacuum tube, which provided the basis for a number of other developments. One was the ability to amplify a sound stimulus, which made possible far more precise audiometers that presented sounds with precisely controlled volume. In addition, the audio oscillator

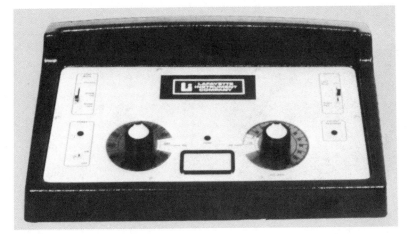

FIGURE 19. Modern solid-state audiometer (reproduced with the permission of the Lafayette Instrument Company, Inc.).

enabled tones to be produced with a high degree of purity and having any desired frequency and intensity (Warren, 1934). In 1942 Boring wrote:

> Thus it came about that the radio age unfrocked Koenig in his sanctum, the psychological laboratory. Nowadays the audio-oscillator replaces the tuning fork. Intensity, being easily controlled, has at last become an important variable

FIGURE 20. Metronome, used in early laboratories as both a timing device and auditory stimulus (photographed by the author).

FIGURE 21. Modern electric metronome (reproduced with the permission of the Lafayette Instrument Company, Inc.).

of the stimulus. The psychologist can now get any frequency at any intensity in the perfect quiet of an electrical potential, changing it at need into sound at the ear-drum by means of ear-phones or a loud-speaker. (p. 332)

Today, psychologists have ready access to precise audiometers such as the one shown in FIGURE 19, but, functionally, much is the same, although there has been somewhat of a shift from using the audiometer for basic research to using it for diagnostic and evaluative purposes.

Even the familiar metronome (FIG. 20), so important in early laboratories both as a timing device and, with its clicking sound, as an auditory stimulus, is still with us in today's instrument catalogues (FIG. 21), as is the tuning fork (FIG. 22).

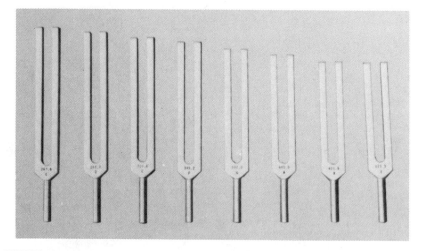

FIGURE 22. A modern series of tuning forks (reproduced with the permission of the Lafayette Instrument Company, Inc.).

INSTRUMENTS FOR THE STUDY OF VISION

If the psychological study of auditory phenomena gave rise to a profusion of instruments, the study of visual processes gave rise to just as many. Some tools were borrowed from the physicist's study of optics, such as the spectroscope, which employed a prism to make the spectrum visible.

The color wheel (FIG. 23) originated as a device with which one could rapidly rotate disks containing two or more colors in order to demonstrate the laws of color mixture, such as the fact that all seven colors of the spectrum will generate white. When the color wheel became a standard device in every

FIGURE 23. An early electric color wheel, used to demonstrate the laws of color mixture (Scripture, 1897).

psychology laboratory, however, the emphasis expanded to include the *psychological* experience of color and the relationship of the physical laws of color mixture to the anatomy and physiology of the eye.

Other visual phenomena generated much interest and thus stimulated the design of apparatus. The study of color contrast led Hering to develop several elaborate pieces of apparatus, such as the Simultaneous Color Contrast Apparatus, which allowed the two eyes to be simultaneously affected by stimuli that differed in color.

In particular, the phenomena of color blindness stimulated much interest because, ultimately, they had to be accounted for by theories of color vision. Devices developed for the diagnosis and study of color blindness were as diverse as Holmgren's wools, with which subjects matched skeins of different-colored yarn with standard skeins (Warren, 1934), and Hering's color blindness apparatus (FIG. 24), which consisted of a tube, through which

FIGURE 24. Hering color blindness apparatus (photographed by the author).

a circular field, half red and half green, was visible. The subject was told to match the two halves of the circular field by moving reflecting screens, and, if he succeeded, he was color blind (Boring, 1950).

Illusions were studied with alacrity, in part because any experience which contradicted the usual veridicality of the senses was viewed with some fascination. Some illusions required only simple line drawings for their demonstration, although other illusions were explored through apparatus. While many such pieces of apparatus originated as games and entertainments, they soon found their way into laboratories for systematic study.

FIGURE 25. One version of the zoetrope, or cylindrical stroboscope (Scripture, 1897).

One illusion that required apparatus for its study was the illusion of perceived movement, which could be demonstrated by the zoetrope or cylindrical stroboscope (FIG. 25). Boring has described the cylindrical stroboscope as follows:

> A strip of paper, bearing pictures showing the successive positions of some object, is slipped inside the cylinder around its wall at the bottom and is viewed from above down through the slits at the top. . . . In this way the stroboscopic principle that the discrete displacement of the stimulus-object can give rise to the perception of continuous movement was fully established. (1942, p. 591)

The development of apparatus for the illusion of perceived movement gave rise to two trends. One was the development of devices for the viewing of long strips of pictures in rapid succession, such as Edison's kinetoscope (FIG. 26).

FIGURE 26. Interior of Edison's kinetoscope (Scripture, 1897).

The kinetoscope held from fifty to one hundred and fifty feet of film in an endless loop. As used in the penny arcades of the time, when a penny was put into the slot, an electric motor started the film moving and each frame was momentarily illuminated as it went by the magnifying eyepiece (Scripture, 1897). Such devices eventually led to the development of motion pictures.

Within the laboratory the zoetrope was a popular piece of apparatus.

Some versions were designed so that their rotation speed and the size of the viewing slits could be regulated and systematically varied (FIG. 27). The somewhat cumbersome zoetrope was eventually replaced by more precise devices such as the tachistoscope with which Wertheimer made his early studies of the Phi Phenomenon (Wertheimer, 1912).

The tachistocope was, of course, a staple instrument for many early studies of visual stimuli since it enabled the presentation of visual stimuli for very short periods of exposure. Tachistoscopes varied widely in complexity and precision. For example, one of the simpler versions, a fall tachistoscope, regulated the speed of exposure of information by means of an aperture falling

FIGURE 27. A motorized form of the cylindrical stroboscope (photographed by the author).

past the information to be displayed. The rate of fall, and thus of exposure, was determined by gravity (see Woodworth, 1938, p. 689).

Although limitations of space prevent a complete discussion of apparatus employed in early studies of vision-related problems, one clear trend is apparent: Many of the early instruments for studying vision are still with us in some form. The color wheel in FIGURE 23 is not significantly different from the contemporary version in FIGURE 28. The Phi Phenomenon has been "packaged" and has its own apparatus now (FIG. 29) and devices abound for the study and demonstration of other illusions. A glance at almost any

FIGURE 28. A modern color mixer (reproduced with the permission of the Lafayette Instrument Company, Inc.).

contemporary instrument maker's catalogue will reveal devices for testing color blindness, measuring color contrast, depth perception, and visual acuity, all problems of interest in early laboratories. Numerous varieties of tachistoscope are available today, but their basic functions remain much as they were in the early days of experimental psychology.

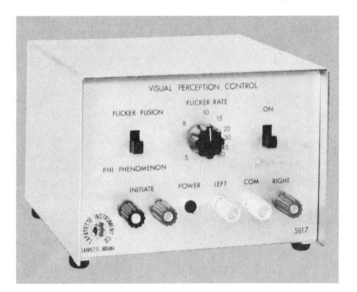

FIGURE 29. Contemporary apparatus used for demonstrating the Phi Phenomenon, or the illusion of perceived movement, as well as for demonstrating the phenomenon of flicker fusion (reproduced with the permission of the Lafayette Instrument Company, Inc.).

MOTOR SKILLS AND LEARNING

In some respects, apparatus for the study of muscle action and coordination provide the most obvious examples of continuity with the past. The knee jerk apparatus for studying reflex action is still very much in evidence in contemporary catalogues, although cosmetic alterations have been made in the design of the equipment. The ergograph, used in experiments on work or fatigue to measure changes in the amount of muscular contraction, has survived relatively unscathed in modern form. In the mirror drawing test, a subject views a design in a mirror and attempts to reproduce it while the hand, pencil, and the actual design are concealed from direct view (Warren, 1934). Versions available for purchase today are almost unchanged from those used during the first decade of the twentieth century. The steadiness apparatus measures involuntary movement as a subject attempts to insert a metal stylus without contact into holes of graduated size. An illustration of this apparatus published in 1910 (FIG. 30; Whipple, 1910) could be incorporated into a modern instrument catalogue since it is almost identical in appearance to its modern counterpart.

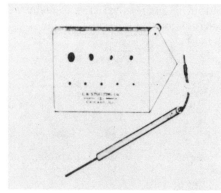

FIGURE 30. An illustration of a steadiness tester (Whipple, 1910) which has changed little in appearance, when compared with contemporary versions.

For the study of memory and serial learning, the elegant Ranschburg memory apparatus (FIG. 31), which rotated a disk containing material to be learned past a small aperture enabling the information to be viewed one item at a time, has been replaced by the sturdier but esthetically less pleasing memory drum (FIG. 32).

In studies of animal learning, there is some continuity between the puzzle boxes of Thorndike, which were never manufactured on a large scale, and mechanisms for the study of operant conditioning (the operant conditioning chamber or "Skinner box") which now flourish in so many laboratories.

Let us not forget to include instances in which apparatus was utilized by our predecessors in ways that were arguably ahead of their time. Let me quote from a recent account of some little-known experiments of John B. Watson,

FIGURE 31. Ranschburg memory apparatus (photographed by the author).

the founder of Behaviorism, who

wanted to know what kinds of biological changes occur in humans during the stress of intercourse. The medical literature in 1917 reported little more than that the pulse rate usually increases—but the hows and whys and whats of the matter were simply not known. Watson tackled the issue directly, by connecting his own body (and that of his female partner) to various scientific instruments

FIGURE 32. Modern memory drum (reproduced with the permission of the Lafayette Instrument Company, Inc.).

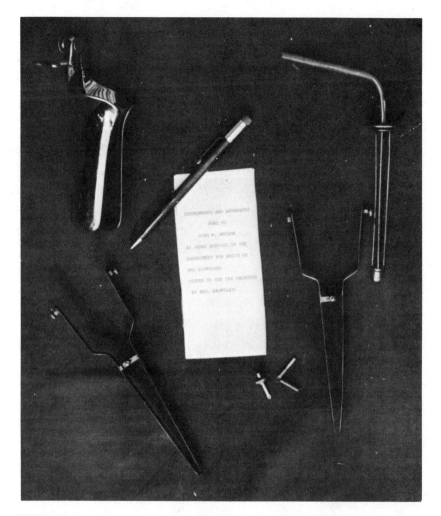

FIGURE 33. Instruments used by John B. Watson in his pioneering studies of human sexual behavior, including a speculum, a sensor thought to have been used for transmitting pressure changes (upper right), and two contacts thought to have been for the purpose of counting contractions (Magoun, 1981). The instruments are now owned by the Canadian Psychological Association Archives at the University of Toronto.

while they made love. He fathered what were probably the very first reliable data on the human sexual response. (McConnell, 1977, p. 273)

Some of the instruments employed by Watson (FIG. 33) have been preserved.[d] Devotees of the history of psychology may be surprised to learn that even the conservative E. B. Titchener must have carried out some form of

[d]Since this paper was presented, a more complete account of Watson's sex research has been published by Magoun, 1981.

research on sexual sensation since he reported in his Text-Book of Psychology that

> both in quality and in its irradiating character the sensation of sexual excitement resembles tickling. We do not know how it is aroused: there are no lust spots, akin to the sensitive spots of the skin. (1910, p. 192)

In this account, Titchener has unfortunately chosen not to describe the methods and apparatus employed to determine that there are no "lust spots." These early efforts of Watson and Titchener to ascertain the nature of sexual sensations and response foreshadowed what has become a major research area in contemporary psychology. Today the researcher can purchase instruments designed expressly to study sexual functions and processes (e.g., FIG. 34). Once again, we see a perhaps unexpected continuity with the past.

INSTRUMENTAL "DINOSAURS"

There are many other areas of experimental psychology where obvious continuities between early apparatus and instruments in use today could be indicated. Gone forever, regrettably, is the sculptural beauty of some early tools, such as a motor employed by Helmholtz (FIG. 35). Vanished is the omnigraph or "spring-driven introspector" (FIG. 36), one attempt to mechanize the circumstances of the process of introspection. (At least, that is what one surmises from its title, although little else is known about this device, which now resides in the Archives of the History of American Psychology at the University of Akron).

No longer do catalogues list the volometer, or "kinetic will test," which was

> devised to test that function of the mind called will, persistency, determination, pluck, or spunk, in terms of muscle fatigue in units of time. (C. H. Stoelting Co., 1936, p. 64)

No account of instruments that have vanished should omit the Psycho-

FIGURE 34. Penile erection feedback system, used to provide information concerning minute changes in penile size (reproduced with the permission of the Stoelting Company).

FIGURE 35. A Helmholtz motor, with a beauty of form and design rarely encountered in contemporary instruments (photographed by the author).

FIGURE 36. Omnigraph, or "spring-driven introspector" (photographed by the author).

graph or Phrenometer. Risse (1976) has given an account of the development of this device designed to measure the size and shape of the skull and to provide evaluations of mental traits according to phrenological principles. It will be recalled that phrenology was a discipline which attempted to correlate specific mental faculties with increased prominence in areas of the skull. According to Risse, the Psychograph was intended to be a diagnostic tool capable of providing vocational guidance to those unemployed as a result of the depression of the 1930s.

In this essay concerning a few of the tools employed by psychologists during the first century of experimental psychology, it is apparent that some instruments have undergone cosmetic and mechanical changes but have retained their basic functions. Others have been discarded, and new instruments have been introduced. One question that has received little attention concerns possible ways in which laboratory apparatus has influenced the course of psychology's development. Have instruments merely been the silent, acquiescent extensions of the psychologists themselves? Or have the instruments themselves in some way played a role in determining the outcomes of laboratory research? Let us look at a few examples of how instruments may have affected outcomes.

POSSIBLE INFLUENCES OF INSTRUMENTS ON EXPERIMENTAL OUTCOMES

For example, consider the problems caused by a tendency to believe in the infallibility of instruments. Early psychology laboratories attempted to imitate the precision of measurement that had been achieved in other sciences. There was an implicit belief that, if psychology followed the examples of physics and physiology and constructed appropriate instruments, almost any problem could be solved. Occasionally, this resulted in what might best be referred to as overconfidence, as in Wundt's initial belief that the Complication Apparatus (FIG. 2) had enabled the discovery of a kind of psychic constant, the time taken by the swiftest thought.

In addition to such errors in evaluating the results of experiments, the mechanical properties of instruments, when incompletely understood, have sometimes been sources of error in measurement. Sokal, Davis, and Merzbach (1976) have noted that

> Any history of the role of the reaction-time experiment in psychology that ignores the Hipp chronoscope and the many problems connected with its use is not only incomplete, but also likely to be inaccurate. (p. 60)

These authors quote from a letter written by James McKeen Cattell in 1884 when he was in Leipzig working with Wundt:

> The trouble is not that one must know physics, but that he must be an original investigator in physics. For example, Professor Wundt thought that when a magnet was made by passing a current around a piece of iron, it was made instantaneously. I find with the current he used it takes over one tenth of a

second. All the times he measured were much too long. Now the time required for magnetization to be developed in soft iron has nothing to do with psychology, yet if I had not spent a great deal of time on this subject, all my work would have been wrong. (p. 60)

Solving the technical and mechanical problems resulting from poorly functioning apparatus has at times required more effort than gathering the actual data. For example, between 1893 and 1931 E. B. Delabarre accumulated a large amount of data concerning his own reactions to *Cannabis indica,* or hashish (Delabarre and Popplestone, 1974). Delabarre's observations included effects of hashish on such physiological functions as pulse, breathing, temperature, muscular steadiness and strength, and measurements of his own reaction times, as well as his performance in tasks such as classification, naming and sorting.

A plaintive note running throughout most of the extensive records he kept concerns his continuing difficulties with apparatus. One example is a note made August 9, 1902:

Have been preparing apparatus for this trial for several years. It has taken two or three years to devise and construct the Synchronous Motor Cylinder Chronograph, and much time also to perfect the Ergograph & Continuous Record Smoked-Paper apparatus, with accessories. . . . Each piece of apparatus has given much trouble and taken much time to put in condition.[e]

Delabarre goes on to describe in detail numerous mechanical changes in the design of the instruments and concludes the day's entry with the comment that "all this has taken very much time. But all was in condition . . . so that it worked perfectly."[e]

Yet, a few days later he notes that he had to make still more repairs, such as repairing the belt of the Ergograph.

Such comments can be found throughout the records of his experimentation, and, although he eventually was forced to suspend much of his research because of problems with apparatus (Delabarre and Popplestone, 1974), his persistence in the face of such constant technical problems was remarkable.

As these examples illustrate, a situation that confronted many psychologists was (and still is) the need to design apparatus for a particular experimental problem, or to modify existing apparatus. This required psychologists to develop skills in the design, construction, and repair of instruments, and either to become engineers themselves or to find those who could carry out their instructions.

Sometimes this was a relatively painless process for the psychologist in question, as in the case of Raymond Dodge, about whom it has been noted that "His inventions are far better known than his contributions of fact and theory." (Yerkes, 1942, p. 590)

[e]E.B. Delabarre papers. Archives of the History of American Psychology, The University of Akron, Akron, Ohio.

For example, in a biographical memoir of Dodge, Walter Miles cited this episode, which led to the creation of Dodge's tachistoscope (FIG. 37):

> [Dodge's] professor in a seminar on the psychology of reading discussed the need for a special piece of apparatus which could serve to exhibit a word or diagram all at once and in clear view for binocular reading or perception. The desirable features of this ideal tachistoscope the professor could enumerate but he could not picture what the apparatus would look like, and he expressed to his seminar the opinion that it would not be possible to build such a piece of equipment. All unwittingly, by this little addendum to his discussion on the psychology of reading, the professor had started a train of scientific events which was to produce a change on the face of Western elementary education. Dodge took the prescription for the tachistoscope which Professor Erdmann had propounded and made it his own problem and project for scientific meditation and planning, and to the astonishment of his professor came up with the answer. (Miles, 1956, pp. 70–71)

Thus, Dodge seems to have had a knack for instrument design. One can only speculate concerning how many would-be experimental psychologists were not so well endowed and were discouraged from pursuing investigations requiring complicated apparatus. I have no data on this, but I can inject an autobiographical note, for, when I was considering various possibilities in the course of selecting a Ph.D. dissertation topic, one problem I considered was in

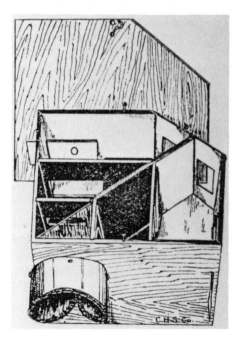

FIGURE 37. Interior of the Dodge tachistoscope (C. H. Stoelting Co., 1936).

the domain of perception. However the very first person I approached as a possible dissertation supervisor turned me down flat because I admitted to having no skills in apparatus construction and wiring and no knowledge of electronics. My aspirations to complete a laboratory dissertation were thoroughly quenched and I eventually completed a thesis in the psychology of creativity which, to my great relief, required no apparatus at all.

Occasionally, engineering problems have theoretically been soluble but have remained unsolved because of the large financial outlay that would be required, and here I will use a much more recent example: apparatus for the study of the Critical Flicker Fusion rate or CFF. Flicker fusion refers to the point at which an intermittent light source appears to a subject to become a steady light source, when the frequency of the intermittent light is controlled by the experimenter. An individual's Critical Flicker Fusion Rate provides a very useful index of the general level of activity of the central nervous system. This is a long-studied phenomenon and was explored in early laboratories by means of the episcotister, a disk with adjustable open and closed sectors that created flicker when it was rotated in front of a steady light source (FIG. 38).

Misiak was the first to publish data on CFF obtained through electronic equipment (Misiak, 1951) and has made a number of significant contributions in describing, for example, developmental changes in CFF (Loranger & Misiak, 1959). He felt, and still believes, that CFF has enormous potential as a research and diagnostic tool that has not been fully realized. For example, Misiak has pointed out that there is evidence that the CFF can detect multiple sclerosis long before the onset of observable symptoms.[f]

However, in order to use CFF as a diagnostic tool it was necessary to establish norms and, in the course of obtaining these data, an unexpected problem arose. As Misiak has described it,

> We found to our surprise that when we put a new bulb in, when a bulb was burned out, the readings were different. It was a question of brightness, which was one of the most important parameters in CFF. The higher the brightness, ... the higher is CFF. These little variations in the bulbs of the same manufacturer would make tremendous differences. Of course we could control the intensity of light but we had to be sure that we had the same intensity from bulb to bulb, from apparatus to apparatus, from situation to situation. [Because of problems in standardizing the intensity of light] we realized that hundreds, thousands, of data were useless.[f] (see also Misiak, 1967).

Misiak had sought to develop a portable CFF unit that could be taken to different locations and persuaded a manufacturer to build such an instrument, which was then used in an ambitious and far-reaching program for the collection of standardization data. However, the problem of light intensity, discovered only after the instrument was manufactured, made it impossible to obtain the precision in data necessary to formulate norms. In addition, there were problems due to variations in the voltage of the local electrical current coming into the apparatus.

[f]Recorded interview with H. Misiak, October, 1980.

FIGURE 38. Episcotister, used in early studies of flicker fusion (Titchener, 1905/1971).

Because of Misiak's pioneering studies with Critical Flicker Fusion and his work in developing the first electronic equipment for its demonstration and study, apparatus for CFF has been incorporated into the standard listing of equipment now available from instrument manufacturers (e.g., FIG. 29). However, as Misiak has pointed out,[f] the technological problems that he and those working with him encountered have still not been fully overcome, with the result that reliable norms remain unavailable, and the potential value of Critical Flicker Fusion as a research and diagnostic tool has yet to be fully realized.

Noting the great dependency upon instrumentation of the psychologist who would address neuropsychological and psychophysical problems, Misiak has urged much closer collaboration between psychologists and engineers, as well as adequate funding for the development of potentially valuable forms of apparatus.[f] This is a fundamental point which those who would build upon and continue Misiak's exciting work—indeed, any psychologist who wishes to address problems of such magnitude and with such potentially valuable pay-offs—would do well to consider.

We have thus far seen a number of ways in which the actual functioning of the apparatus may have affected the study of psychological processes. However, there is another, broad way in which apparatus can affect outcomes, and this is with respect to how the availability and current popularity of particular types of apparatus influence the choices of experimental problems.

Lubeck (1979) has made an analysis of patterns of research on aggression in social psychology. He has pointed out that after the pioneering work of Dollard, there was a renewal of interest in conducting aggression research in the laboratory. Lubeck notes that one factor which contributed to this interest was the development of the "aggression machine," or "shock-box." He points out that

> Since about 1960, measures of intensity, duration, frequency, and/or latency of delivery of electric shock (usually pseudoshock) to another S (usually a pseudo-S) have increasingly been employed deceptively in studies involving, for the most part, college students or adults. (p. 291)

In a survey of journal articles that employed electric-shock delivery as a dependent variable, Lubeck reports finding essentially linear trends from 1956 to 1977, with a clear correspondence between the number of aggression articles and the percentage of those articles which used the "shock-box" method. He interprets these data as indicating that "the delivery of electric shock seemed well accepted, until perhaps just recently, as *the* method in sociopsychological aggression research." (p. 291)

Lubeck's findings suggest that a fertile medium for exploring the social psychology of methodology selection might be the frequency, over time, with which a particular piece of apparatus is selected for work within a general problem area.

The variety of apparatus available for purchase may also affect choices of problems, and such availability from manufacturers is to some extent dependent upon continued profitability. For example, the 1936 catalogue of C. H. Stoelting Co. is some 227 pages in length, with a six-page index listing approximately 1200 items, when individual pieces, variations in models, and accessories are considered. By 1978, the Stoelting catalog was 46 pages in length, with a four-page index listing approximately 750 items and listing as well some 21 items that had been discontinued from the previous year, presumably because they were no longer profitable or in demand. Once equipment is no longer manufactured, the possibilities for research are reduced and future experimenters wishing to explore additional aspects of older problems may then be faced with the problems of designing and constructing apparatus mentioned earlier.

In general, the popularity of those pieces of apparatus currently in vogue is reflected in greater varieties of models. For example, for the Operant Conditioning Chamber (or "Skinner Box") there are standard and deluxe models, each with numerous combinations of accessories.

I have yet to carry out a survey of availability of this apparatus and its variations in instrument catalogues since its introduction by Skinner, but I believe that such an analysis would mirror fairly closely the impact of operant conditioning paradigms on recent historical trends in methodology employed in laboratory psychology. It is not out of the question that, just as the recognition of the phenomenon of experimenter demand caused great consternation in academic circles during the 1960s, so also might future historians discover that a kind of "apparatus demand" has sometimes influenced the choices of problems, the training of students, the distribution of financial resources through the heavier funding of grants specifying particular methodologies, and the acceptance of findings for publication in scholarly journals.

It seems fitting to close by returning to Wundt in his hypothetical visit to a modern laboratory and to ask him, "Is this what you expected, Herr Wundt? Have we succeeded? Is there now a science of psychology?"

I will not presume to speak for Wundt and will leave you to judge how he might answer those questions, were he here today. However, it seems reasonable to suppose that, with the hindsight of the past 100 years to draw upon, Wundt might suggest that the twentieth-century psychologist needs to consider more carefully the influence of apparatus in the formulation of questions that will be explored during the next 100 years of experimental psychology.

SUMMARY

This paper explores aspects of the role that laboratory instruments have played in establishing psychology as a science. Comparisons of earlier instruments with their more modern counterparts indicate that many problems and methods still of interest today were explored in the laboratories of psychology's pioneers. Such continuities are noted for apparatus employed in the study of reaction time, studies of auditory and visual processes, motor skills, and learning. Several examples are noted of instruments no longer in use, and a discussion is provided of ways in which instruments themselves may influence the outcome of experimental research.

ACKNOWLEDGMENTS

I would like to express my appreciation to John A. Popplestone, Director, and Marion White McPherson, Associate Director, of the Archives of the History of American Psychology, The University of Akron, and Cedric A. Larson of Rutgers–The State University (retired) for their assistance in locating several of the sources utilized in this paper. I would also like to thank Henryk Misiak, Professor Emeritus of Psychology, Fordham University, for sharing with me his insights concerning the role of laboratory apparatus in his research concerning Critical Flicker Fusion.

I would also like to thank members of my family for their assistance, with special thanks to Mrs. Charles N. Ingram and Mrs. Lewis Vickers, Jr. Their help has been invaluable to me.

REFERENCES

BORING, E. G. 1942. Sensation and Perception in the History of Experimental Psychology. Appleton-Century. New York.

BORING, E. G. 1950. A History of Experimental Psychology (2nd edit.). Appleton-Century-Crofts. New York.

BRINGMANN, W. G., G. A. UNGERER & H. GANZER. 1980. Illustrations from the life and work of Wilhelm Wundt. *In* Wundt Studies: A Centennial Collection. W. G. Bringmann & R. D. Tweney, Eds., C. J. Hogrefe. Toronto.

C. H. STOELTING CO. 1936. Psychological and Physiological Apparatus and Supplies. C. H. Stoelting Co. Chicago.

C. H. STOELTING CO. 1978. 1978 Catalog of Experimental Psychology Instruments. C. H. Stoelting Co. Chicago.

DAVIS, A. B. & U. C. MERZBACH. 1975. Early Auditory Studies: Activities in the Psychology Laboratories of American Universities. Smithsonian Institution Press. Washington, D.C. (U.S. Government Printing Office Document Number 047-001-00124-9)

DELABARRE, E. B. & J. A. POPPLESTONE. 1974. A cross cultural contribution to the Cannabis experience. The Psychological Record 24: 67–73.

DIAMOND, S. 1974. The Roots of Psychology: A Sourcebook in the History of Ideas. Basic Books. New York.

GALTON, F. 1883. Inquiries into Human Faculty and Its Development. Dent. London.

HELMHOLTZ, H. L. F. [On the Sensations of Tone as a Physiological Basis for the Theory of Music.] (A. J. Ellis, Ed. and trans.). Dover Publications. New York. 1954. (Originally published, 1863 and translated into English in 1885).

JAMES, W. Letter to Thomas W. Ward, ca. November, 1867. *In* The Letters of William James, Vol. 1. H. James, Ed. Atlantic Monthly Press. Boston. 1920. pp. 118–119.

LORANGER, A. W. & H. MISIAK. 1959. Critical flicker frequency and some intellectual functions in old age. Journal of Gerontology 14 (3): 323–327.

LUBECK, I. 1979. A brief social psychological analysis of research on aggression in social psychology. *In* Psychology in Social Context. A. R. Buss, Ed. John Wiley & Sons. New York.

MAGOUN, H. W. 1981. John B. Watson and the study of human sexual behavior. The Journal of Sex Research 17: 368–378.

MCCONNELL, J. V. 1977. Understanding Human Behavior (2nd edit.). Holt, Rinehart & Winston. New York.

MILES, W. R. 1956. Raymond Dodge, 1871–1942: A Biographical Memoir. Columbia University Press. New York. Published for the National Academy of Sciences and reprinted from Biographical Memoirs, Vol. 29: 65–122.

MISIAK, H. 1951. The decrease of critical flicker frequency with age. Science 113 (No. 2941): 551–552.

MISIAK, H. 1967. The flicker-fusion test and its applications. Transactions of The New York Academy of Sciences, Series II, Vol. 29 (No. 5): 616–622.

RISSE, G. B. 1976. Vocational guidance during the depression: Phrenology versus applied psychology. Journal of the History of the Behavioral Sciences 12 (2): 130–140.

SCRIPTURE, E. W. 1895. Thinking, Feeling, Doing. Chautauqua-Century Press. New York.

SCRIPTURE, E. W. 1897. The New Psychology. Charles Scribner's Sons. New York.

SOKAL, M. M., A. B. DAVIS & U. C. MERZBACH. 1976. Laboratory instruments in the history of psychology. Journal of the History of the Behavioral Sciences, 12 (1): 59–64.

TITCHNER, E. B. 1905. Experimental Psychology: A Manual of Laboratory Practice. Volume II, Quantitative Experiments; Part II, Instructor's Manual. Macmillan. New York. (Reprinted 1971, Johnson Reprint Corporation. New York.)

TITCHENER, E. B. 1910. A Text-Book of Psychology. Macmillan. New York. (Reprinted, 1916)

WARREN, H. C. 1934. Dictionary of Psychology. Houghton Mifflin Company. New York.

WERTHEIMER, M. 1912. Max Wertheimer (1880–1943) on the phi phenomenon as an example of nativism in perception, 1912. Excerpt from Wertheimer, M. Experimentelle Studien über das Sehen von Bewegung, Zeitschrift für Psychologie, 61.Translated by D. Cantor. In A Source Book in the History of Psychology. R. J. Herrnstein & E. G. Boring, Eds., Harvard University Press. Cambridge, MA. 1965.

WHIPPLE, G. M. 1910. Manual of Mental and Physical Tests. Warwick & York. Baltimore.

WOODWORTH, R. S. 1938. Experimental Psychology. Henry Holt and Company. New York.

YERKES, R. M. 1942. Raymond Dodge: 1871–1942. American Journal of Psychology 55: 584–600.

Mathematical Probability and the Reasonable Man of the Eighteenth Century[a,1]

LORRAINE J. DASTON

Department of the History of Science
Harvard University
Cambridge, Massachusetts 02138

INTRODUCTION

The intellectual historian's stock-in-trade consists in tracing the filiation and cross-fertilization of ideas, showing the hybridization of key concepts among disciplines. Often our assessment of the plausibility of one such historical reconstruction over another depends on our mental map of the disciplines involved, which specifies their intellectual proximity in some way. Thus we are not surprised to discover a flow of influences between, say, physics and astronomy, or between sociology and law, for these rank as near neighbors on our disciplinary chart. But a claim for the conceptual interaction of astronomy and law would raise eyebrows. We also rank the relative permeability of disciplinary boundaries when evaluating the case for a purported borrowing of ideas: hence anthropology borrowing from thermodynamics seems more immediately credible than the reverse, for we commonly judge thermodynamics to rank higher within the hierarchy of conceptual autonomy than anthropology. (This hierarchy corresponds closely to our ordering of scientific disciplines by degrees of "hardness" and "softness.")

Along both of these dimensions of proximity and permeability mathematics occupies a rather insulated position. Although we credit an account of occasional loans from philosophy and inspiration from physics, even close disciplinary neighbors like the latter tend to be regarded as the beneficiaries rather than as the donors of ideas when mathematics is involved. No doubt this picture is largely accurate. However, we err when we attempt to impose contemporary maps of the intellectual terrain upon the past without modification. Disciplines emerge, disappear, and shift position with respect to one another like continents, and at a far more rapid pace. Disciplines now perceived as remote may once have bordered upon one another, and conversely, then apparently disparate subjects may now be seen as closely allied. A careful study of this disciplinary drift may serve to illuminate otherwise puzzling episodes in intellectual history. My topic, the relationship between classical probability theory and the moral sciences—as the study of man and

[a]This paper was delivered to the Section of History, Philosophy and Ethical Issues of Science and Technology of The New York Academy of Sciences on 28 May 1980.

0077–8923/83/0412–0057 $01.75/2 © 1983, NYAS

society was termed in the eighteenth century—is meant as a case in point. I will argue that several salient aspects of the development of mathematical probability during the eighteenth century can only be understood in relation to attempts to create a science of rational choice and conduct. In particular, the central concept of probabilistic expectation underwent several striking changes as mathematicians strove to capture the essence of rational belief and action in their definition of expectation.

My argument will consist of several parts: first, I will show that expectation, rather than probability *per se* was the fundamental concept in the earliest formulations of mathematical probability and that it continued to play an important role in eighteenth-century expositions of the theory; second, that probabilistic expectation inherited two distinct qualitative senses of expectation, the one derived from legal notions of equity and the other from economic considerations of prudence, which made expectation the bridge between mathematical probability and the moral sciences; and third, that the program for "mixed mathematics" obliged mathematicians to adapt and amend the theory of probability to tally with reigning conceptions in the moral sciences, in particular that of an archetypal "reasonable man." I will discuss in some detail how this mandate to match mathematical theory to the accepted understanding of rationality compelled mathematicians to successively modify their definition of probabilistic expectation, justifying it first on legal, then on economic, and finally upon psychological grounds. Only with the advent of a new model of explanation for the social sciences that emphasized social regularities rather than individual rationality, coupled with the recognition of the independence of mathematical probability from its applications, did probabilistic expectation cease to be a matter of mathematical controversy (although economists continue to debate closely related questions in so-called rational choice theory). My aim is not only to explain the at first glance surprising relationship between mathematical probability and the moral sciences during this period, but also to shed some light on the problem of applying mathematics to experience as it was conceived in the eighteenth century.

CLASSICAL PROBABILITY THEORY AND EXPECTATION

By almost all accounts, mathematical probability theory originated with the correspondence between Pascal and Fermat in 1654, although others had earlier speculated on the mathematics of gambling. The true originality of the Pascal-Fermat exchange lay neither in their choice of problems—the so-called problem of points had appeared in the mathematical literature as early as 1495—nor in the enumeration of all possible combinations, which had also appeared earlier, but rather in the concept of equal expectation and equiprobability. Both mathematicians viewed the problem as one of determining expectations rather than probabilities.[2]

Although Pascal and Fermat invented the calculus of probabilities in their 1654 correspondence, their letters remained unpublished until 1679. Christian Huygens, visiting Paris in 1655, heard about the problem and the exchange from mutual friends and reconstructed the solutions. After confirming his answers with the French mathematicians, Huygens composed a brief treatise on the subject, which was published in Latin in 1657, and subsequently translated into French, Dutch, and English. Huygens' treatise was not only the first published work on mathematical probability, it was also the first to organize the theory into a system of definitions, postulates (termed "hypotheses"), and propositions.[3] His approach dominated the subject until the posthumous publication of Jakob Bernoulli's *Ars conjectandi* in 1713. And even in Bernoulli's work Huygens' treatise was taken as a departure point; it was reprinted with commentary as Book I of Bernoulli's more comprehensive treatment of the subject. Huygens' influence on classical probability theory was therefore both seminal and enduring, and his approach can be legitimately taken as characteristic.[4]

Strictly speaking, Huygens' treatise set forth a calculus of expectations rather than of probabilities. Huygens posed problems on the fair division of stakes or the "reasonable" price for a player's place in an ongoing game, rather than questions about the probabilities of the events themselves. Considered by itself, Huygens' fundamental principle—his definition of expectation—sounds suspiciously circular:

> I begin with the hypothesis that in a game the chance one has to win something has a value such that if one possessed this value, one could procure the same chance in an equitable game, that is in a game that works to no one's disadvantage.[5]

Since later probabilists *defined* an equitable or fair game as one in which the players' expectations equaled the price of playing the game (the stake), Huygens' definition seems to lead nowhere. In later mathematical parlance, expectation would be defined as the product of the probability of an outcome and the value of that outcome. (For example, the expectation of a player holding one out of 1000 tickets in a fair lottery with a prize of $10,000 would be

$$1/1000 \times \$10,000 = \$10,$$

the fair price of a single ticket.) However, Huygens here assumed that the notion of an equitable game was a self-evident one for his readers. The alternative definition, which gained currency in the eighteenth century, derived expectation and the criterion for a fair game *from* the anterior definition of probability, expressed as the ratio of the number of combinations favorable to an event to the total number of combinations. This route remained closed to Huygens. Instead, he appealed to an intuitive, or at least non-mathematical, notion of equity: in this case, the equitable exchange of expectations and the conditions for a fair game.

Why did Huygens and the first generation of mathematical probabilists choose expectation rather than probability proper as the foundation for their theory? In part because readers could be presumed familiar with the qualitative legal analogue of expectation. Within the corpus of Roman and canon law, aleatory contracts—that is, all agreements involving an element of chance—constituted a recognized subdivision of contract law including, for example, games of chance, annuities, partnerships in which one party supplied the capital and another assumed the burden of the risk of the venture, maritime insurance, prior purchase of a "cast of a net" of a fisherman. This type of contract attracted much attention from sixteenth- and seventeenth-century mercantile apologists who hoped to class loans with interest and other apparently usurious practices under this more innocuous heading within the law.

As in all contract law of the period, jurists were primarily concerned to stipulate the conditions of equity between partners to such aleatory contracts. Contracts, according to the famous seventeenth-century jurist Grotius, "were intended to promote a beneficial intercourse among mankind" and therefore presumed equality of terms. Equal expectations, rather than equal probabilities of gain or loss, generally assured an equitable contract, although some jurists also insisted upon an equality among partners, particularly in commercial ventures.[6] But most were willing to accept an agreement in which one party bought, for example, a prospect with only a slim probability of coming to pass for a substantial price, if the value of the prospect were high enough.

Nicholas Bernoulli, nephew and editor of Jakob, summarized the conventional doctrine of legal expectation as consisting "in the right of gaining and having that which will be taken. Thus, the buyer cannot complain of injury if nothing should be taken, since from the beginning gain and loss are quite evenly balanced against one another."[7]

It is significant that the earliest expositions of mathematical probability, as well as the problems addressed by its practitioners, were all couched in terms of expectation, and of expectation conceived as equity. Games of chance were included among aleatory contracts, and the problem of points that had prompted the Pascal-Fermat correspondence was a typical problem in legal expectation: how to divide the stakes of an unfinished game fairly. The determination of the just stake in a game of chance was another. Huygens' above-quoted definition of expectation was cast in terms of a fair trade or contract; equal expectations were those which could be exchanged for one another "in a just and equal game." Several of Huygens' demonstrations hinge on an intuitive understanding of an equitable contract. In order to prove expectations equal, Huygens posited a series of "equitable" trades among players. Later probabilists defined expectations in terms of probabilities; Huygens derived probabilities (implicitly) from expectations. Chances are equiprobable *because* the game is assumed fair.

Expectation-based treatments soon gave way to more genuinely probabilistic ones. By 1709, Nicholas Bernoulli could reproach his fellow jurists for their imprecise understanding of expectation, holding up the mathematical "art of conjecture" as a surer guide to equity in aleatory contracts, particularly annuities.[8] Probabilistic expectation defined the terms for an equitable contract, rather than the reverse, in this new scheme. However, Nicholas Bernoulli upheld the older identification of equal expectation with equitable contracts and exchanges; he simply proposed to substitute computation for more intuitive reasoning. For continental probabilists, expectation retained its legal overtones of equity until the end of the eighteenth century.

EXPECTATION AND ECONOMIC SELF-INTEREST

Probabilistic expectation also came to be influenced by a more purely economic brand of reasoning in late seventeenth-century philosophical and religious discussions of rational belief under conditions of incorrigible uncertainty. Skeptical attacks on the rationalist ideal of demonstrative knowledge had made considerable headway in philosophical circles, and the new model of empirical, tentative proof exemplified by Newton's celebrated methodological reflections in the General Scholium of the *Principia* had further weakened the deductive program for science. Many echoed Locke's contention that the majority of human decisions must be made in the "twilight of probabilities" rather than in the noonday glare of certainty, but still hoped to steer a middle course between the nihilism of the pyrrhonists and the rigidity of the School and their rationalist successors. Prominent among these moderates were the so-called Royal Society theologians, Robert Boyle, John Wilkins, and Joseph Glanvill.[9] While conceding that mathematical or "metaphysical" certainty might elude the grasp of merely human intellects without the aid of revelation, these apologists nonetheless maintained that rational belief—religious, scientific, or other—was justified because daily life would be unthinkable without it. Belief, they argued, was practical and active as well as intellectual and contemplative. By this pragmatic criterion, even the most confirmed skeptic suspended his disbelief long enough to take meals, thus betraying by his actions a stubborn will to believe that belied his professed philosophical doubts in the existence of an external world. In this vein, Boyle observed that although moral demonstrations based on a "concurrence of probabilities" could not lay claim to metaphysical or even the lesser degree of physical certainty, they were "still the surest guide, which the actions of men, though not their contemplations, have regularly allowed them to follow."[10]

The apologists took this "practical reason" of everyday life to be their standard in all belief: we are obliged to believe whatever is sufficiently likely, be it the law of gravitation, the existence of God, or the permanence of taxes, as to persuade a reasonable man to take action in the ordinary course of his affairs. By "probable," the apologists generally meant "highest expectation." The actual probability of God's existence, or the success of a voyage to the East Indies, must be weighed against the magnitude of the possible gain or loss. Pascal's wager was the paradigm case for such reasoning by expectation, and enjoyed wide currency both in England and in France during the latter half of the seventeenth century.

The "thoroughly prudent man" of Pascal's wager made appearances in many religious tracts of this period. Boyle claimed that the "dictates of prudence" endorsed reasoning by expectation: all reasonable men would concur that it was best to sacrifice a gangrene-infected limb in the hopes of saving a life; to submit to unproven remedies for smallpox and other dread diseases when stricken; to invest in a risky commercial venture with a huge prospect of gain.[11] The last instance became the paradigm case for all such prudential arguments. Borrowing a leaf from merchant investors, Boyle and others urged their readers to maximize their expectation in every aspect of their lives, from commerce to religion.

These two perspectives on expectation, the legal and the economic, the one concerned with equity and the other with profit, both shaped the nascent mathematical theory of probability from its earliest formulation in 1654 until the publication of Laplace's *Théorie analytique des probabilités* in 1812. Although the dictates of legal equity and economic prudence tugged the mathematical concepts in different directions, both linked probability theory to Enlightenment views of moral rationality and the human sciences. Seventeenth-century contract theories as elaborated in the works of Grotius, Hobbes, Pufendorf, and Locke, on the one hand, and the new respectability of economic interests in political theory as both a check to more dangerous vices and a goad to belief and action on the other, were two prominent themes in the moral sciences of the period.[12] The rival definitions of expectation advanced by mathematicians during this period show it to have been a pressure point for the influences of social theory, responding first to the proponents of equity, then to those of prudence.

MIXED MATHEMATICS AND THE MORAL SCIENCES

The eighteenth-century moral sciences correspond only approximately to the latter day social sciences, although the two share many common concerns and are of course historically continuous. For the purposes of this discussion, the two salient points of contrast concern objectives and units of analysis. The moral sciences sought not only to formulate theories that would describe (and ideally predict) psychological and social phenomena but also undertook to establish standards for rational thought and conduct. Descriptive and prescriptive elements were so closely intertwined as to be inseparable. Even the Physiocrats, who professed to seek the "natural laws" that governed the social realm, understood these laws in a very different way than that in which, for example, a physicist might. Obedience to physical laws was not a matter of choice, but submission to the laws of the moral realm was voluntary, although in the best interests of the individual and of society. For Comte, Quetelet, and other nineteenth-century social theorists, in contrast, the laws of pyschology, sociology, and economics were as inexorable as those of physics. The eighteenth-century moral sciences studied the deliberations and behavior of a select group of individuals designated as "rational" (*hommes éclairés*) in the hopes of deriving a set of explicit rules to guide the less astute majority. Their perspective was psychological and individualistic, as well as prescriptive. Societies figured in these theories only as aggregates of individuals, with properties inferred from the sum of their parts. In contrast, nineteenth-century social theorists investigated societies as coherent units (Comte went so far as to deny psychology the status of an independent science), and expected to discover law-like regularities only at the macroscopic level.

To its eighteenth-century practitioners, classical probability seemed the uniquely appropriate mathematical tool for the analysis of the thought processes of the rational individual investigated by the moral sciences. By codifying the principles that guided an elite of reasonable men, the probabilists hoped to make rationality accessible to all, for as Voltaire quipped, common sense was not all that common. When the results of mathematical probability contradicted the judgments of these *hommes éclairés,* mathematicians sought to realign the mathematical results with enlightened opinion. Probabilists insisted that the mathematical theory only described and systematized, rather than dictated, reasonableness.

The debate over the conception of probabilistic expectation hinged on this elusive idea of "reasonableness." Throughout the discussions of the probabilists, the key notion of the "reasonable man" who comports himself so sensibly in mundane matters is left undefined. His judgment is made the measure of all rational belief and conduct, but the nature and validity of that judgment is taken as given. The meaning of reasonable conduct was stipulated as both self-evident and fixed, the permanent standard for all decisions. However, the accepted sense of "reasonable" did undergo changes during this period, and the history of the debate over the definition of probabilistic expectation is that of the mathematicians' attempts to keep pace with these shifts.

Before documenting these influences, one must ask the prior question, in light of the above observations on the peculiar "impermeability" of mathematics: why was probability theory so sensitive to these extra-mathematical pressures? Or, posed somewhat more pointedly, why did probability theory seem the best candidate for a "social mathematics," in Condorcet's phrase? Understood in the context of eighteenth-century views on the nature and role of "mixed mathematics," these questions turn out to be the reverse and obverse of the same issue. Examining the closest thing we have to a map of the intellectual terrain of the Enlightenment, the chart of the "System of Human Understanding" appended to d'Alembert's *Preliminary Discourse* (1751) of the *Encyclopédie,* we find that mixed mathematics dominates the category of mathematics. Pure mathematics includes only geometry and arithmetic (of which algebra and the calculus are subdivisions), but mixed mathematics embraces mechanics, optics, acoustics, "geometric astronomy," pneumatics— in short, all of the exact sciences, plus the "art of conjecture" (Jakob Bernoulli's term for mathematical probability theory, after the *Port-Royal Logic, or Art of Thinking*). Situated midway on the chart between pure mathematics and "particular physics" (zoology, botany, meteorology, chemistry), mixed mathematics bridged mathematics proper and the natural sciences.

It would be tempting to equate mixed mathematics with contemporary applied mathematics, but the temptation should be resisted. Pure and mixed mathematics are related as contiguous sections of a continuum; pure and applied mathematics as two distinct levels in a hierarchy. Eighteenth-century mathematicians had every reason to expect a coincidence between mathematics and phenomena, for they subscribed to a Lockean theory of knowledge that derived mathematics *from* phenomena. The pure mathematics of arithmetic, analysis, and geometry was ultimately abstracted from perceptions of the physical world and constituted the most fundamental (and simplest) of the empirical sciences, the most schematic rendering of the phenomena shorn of all features except magnitude and extension. Mixed mathematics represented an intermediate point between the skeletal renditions of pure mathematics and the "blooming, buzzing confusion" of experience from which these were ultimately drawn. As might be expected, the boundaries between pure and mixed mathematics were sometimes blurred: mechanics, for example, might be equally well classed in either category, and was.

As a part of the natural sciences, mixed mathematics consisted in theories about various features of the physical world. Unlike the theories of particular physics in, say, medicine or meteorology, these theories were quantitative ones. However, both types of theories shared an obligation to describe phenomena accurately. For mixed mathematicians, this generally meant achieving near-congruence between the significant aspects of a given class of phenomena and the mathematical model that described it. If the mathematical description diverged palpably from the phenomena, it was incumbent upon the mixed mathematicians to revise their theory. In this respect, mathematical probability was as corrigible as the mathematical theory of lunar motion. Eighteenth-century probabilists understood their theory as a mathematical description of "reasonableness," subject to the empirical check of the actual opinions and conduct of enlightened men.

Expectation, and with it the calculus of probabilities, thus became bound up with common, albeit equivocal, reason. Laplace concluded his *Essai philosophique sur les probabilités* (1814) by underscoring this connection:

> It is seen in this essay that the theory of probabilities is at bottom only common sense reduced to a calculus; it makes us appreciate with exactitude that which exact minds feel by a sort of instinct without being able ofttimes to give a reason for it.[13]

When mathematical results clashed with good sense, eighteenth-century probabilists anxiously re-examined their premises and demonstrations for errors and inconsistencies.

THE ST. PETERSBURG PARADOX AND THE DEBATE OVER EXPECTATION

This is why the St. Petersburg paradox, trivial in itself, triggered an animated debate over the foundations of probability theory. Unlike most mathematical paradoxes, the contradiction lay not between discrepant mathematical results reached by methods and assumptions of apparently equal validity, but rather between the unambiguous mathematical solution and good sense. The problem, first proposed by Nicholas Bernoulli in a letter to Pierre Montmort and published in the second edition of the latter's *Essai d'analyse sur les jeux de hazard* (1713), belonged to the staple category of expectation problems. Two players, A and B, play a coin-toss game. If the coin turns up heads on the first toss, B gives A 1 ducat; if heads does not turn up until the second toss, B pays A 2 ducats, and so on, such that if heads does not occur until the nth toss, A wins 2^{n-1} ducats from B. How much should A pay B to play the game? Figured according to the standard definition of expectation, the product of the probability of the outcome and the value of the outcome, A's expectation is infinite, for there is a finite, though vanishingly small probability that even a fair coin will produce an unbroken string of tails. Therefore, A should pay an infinite amount to play the game:

$$E = 1/2(1) + 1/4(2) + 1/8(4) + \cdots + 1/2^n(2^{n-1}) + \cdots$$

However, as Nicholas Bernoulli and subsequent commentators were quick to point out, no reasonable man would pay even a small amount, much less a very large or infinite sum, to play such a game. The results of the standard mathematical analysis clearly affronted common sense; hence the "paradox." The divergent solutions proposed to this dilemma by eighteenth-century probabilists reflected the tension between the equitable and prudential connotations of expectation.

In 1738, Daniel Bernoulli, cousin of Nicholas, published a solution to the paradox in the annals of the Academy of St. Petersburg. Historians of economic theory regard this memoir as the earliest expression of the concept of economic utility, but Bernoulli and his colleagues considered the memoir to be an important contribution to the mathematical theory of probability.[14] Bernoulli's memoir did more than rechristen the problem in honor of the Academy: it also set the acceptable terms of solution for his successors. Although other mathematicians challenged the specifics of Bernoulli's approach, all agreed that the paradox was a real one that threatened to undermine the calculus of probabilities at its foundations; that the definition of expectation was the nub of the problem; and that a satisfactory solution must realign the mathematical theory with the views of reasonable men.

Bernoulli's strategy was to distinguish two senses of expectation, one "mathematical" and the other "moral." Mathematical expectation corresponded to the classical definition of expectation, and Bernoulli's discussion of this type of expectation reflected its legal origins. He observed that the standard definition ignored the individual characteristics of the risk-takers, and that it was premised on the equitable assumption that everyone encountering identical risks deserved equal prospect of having his "desires more closely fulfilled." Bernoulli thus linked "mathematical" expectation with the legal context of aleatory contracts, balancing uncertain expectations against an immediate and certain amount. Mathematical expectation quantified the jurists' intuitive equity obtaining among parties to such contracts, retaining the vocabulary and aims of contract law.

In his treatment of the St. Petersburg problem Bernoulli proposed to shift the perspective from that of a "judgment" of equity pronounced "by the highest judge established by public authority" to one of "deliberation" by an individual contemplating a risk according to his "specific financial circumstances." Once the concept of expectation was transplanted from a legal to an economic framework, the classical definition lost its relevance. Whereas mathematical expectation had been explicitly defined to exclude personal circumstances that might prejudice the judicial assumption of the equal status of contracting parties, fiscal prudence required some consideration of just such specifics. Bernoulli argued that the plight of a poor man holding a lottery ticket with a $\frac{1}{2}$ probability of winning 20,000 ducats was in no way equivalent to that of a rich man in the same position: the poor man would be foolish not to sell his ticket for 9,000 ducats, although his mathematical expectation was 10,000; the rich man would be ill-advised not to buy it for the same amount.

Bernoulli maintained that a new sort of "moral" expectation must be applied to such cases in order to bring the notion of value (i.e., possible gain or loss) into line with prudent practice. Mathematical expectation quite rightly equated outcome value with price, a method well-suited to civil adjudication because it is intrinsic to the object and uniform for everyone. Moral expectation, in contrast, based value on the "utility"—or "the power of a thing to procure us felicity" in the words of the eighteenth-century economic theorist Galiani—yielded by each outcome, which may vary from person to person. Moral expectation was the "mean utility" (*emolumentum medium*), or product of the utility of the outcome and its probability.[15]

In order to estimate utility, Bernoulli supposed as the most general hypothesis that infinitesimal increments of utility were directly proportional to infinitesimal increments in wealth and inversely proportional to the amount of the original fortune. Thus, the function relating utility to actual wealth would be logarithmic. In order to apply utility to real situations, Bernoulli was obliged to incorporate a theory of value into his analysis. He defined wealth as "anything that can contribute to the adequate satisfaction of any sort of want," including luxuries. Bernoulli also invoked a possessive theory of labor, one in which human labor counted as a commodity or possession, to be bought and sold at a market price.[16] In order for the utility function to be everywhere defined, the actual fortune must always be greater than zero:

$$dy = b(dx/x), \text{ where } dy = \text{increment in utility}$$

$$x = \text{actual wealth}$$

$$b = \text{constant of proportionality.}$$

In order that x be always positive, Bernoulli declared that there was "nobody who can be said to possess nothing at all in this sense unless he starves to death. For the great majority the most valuable of their possessions so defined will consist in their productive capacity." Applied to the St. Petersburg problem, moral expectation reduced the stake to a modest sum, affirming the reasonable man's reluctance to play the game except for trifling stakes.

The French mathematician d'Alembert also argued that probability theory must explain prudent action in uncertain situations, but challenged Daniel Bernoulli's solution to the St. Petersburg problem on the grounds that moral expectation, while more accurate than mathematical expectation, still over-simplified the actual experience of risk-taking. D'Alembert accepted Bernoulli's premise that "moral considerations, relative to either the fortune of the players, their circumstances, or even their strength" refined the results of probability theory, but despaired of quantifying all these factors, or even or ordering their relative importance.

Moreover, d'Alembert remarked that Bernoulli had failed to specify the rules for determining whether mathematical or moral expectation applied in a given case. Since ordinary computations of expectation appeared to give satisfactory results in most cases without any consideration of the financial status of the risk-takers, Bernoulli's introduction of moral expectation for the St. Petersburg problem seemed suspiciously *ad hoc* to d'Alembert.[17] For d'Alembert, the St. Petersburg paradox arose from a betrayal of good sense on a different front. Whereas Bernoulli believed that the value term of mathematical expectation overestimated the gains, d'Alembert countered that it was the probability term that absurdly inflated the summed expectation. Although it might be "metaphysically" possible for a fair coin to turn up tails 100, 1000, or *n* times in a row, experience dismissed such outcomes as "physically" impossible. Should such a run of consecutive tails actually occur, d'Alembert claimed that observers would rightly posit some underlying, uniform cause, such as an asymmetric coin. Mathematical probability, which based the postulate of equiprobable outcomes on our ignorance of any cause that might tip the balance in favor of heads or tails, would therefore no longer apply to the situation.

D'Alembert often returned to the vexed notion of expectation in his critical discussions of mathematical probability. Although dissatisfied with Bernoulli's alternative to classical expectation, d'Alembert acknowledged its failure to capture reasonable conduct, even in simple gambling situations. For example, a lottery with an enormous prize, say a million francs, but only a tiny chance of winning, say .0001, would offer an attractive expectation of 100 francs for each ticket according to the classical formula. Yet d'Alembert felt that prudence would counsel against such an investment. For d'Alembert, this discrepancy between mathematical expectation and prudence pointed to serious flaws in probability theory, which had failed to describe reasonable conduct accurately.

As the debate between Daniel Bernoulli and d'Alembert revealed, legal, economic, and even physical strands were woven into the concept of reasonableness. Georges Leclerc Buffon teased out yet another strand, this time a psychological one, in his "Essai d'arithmétique morale" in the Supplement to his *Histoire naturelle*. Buffon described a graduated scale of certainties, progressing from the "purely intellectual" truths of mathematics to physical certainty, founded on an overwhelming mass of evidence, to the far weaker probability of moral certainty, based on analogical reasoning. In the case of the lottery example cited by d'Alembert, Buffon maintained that the probability .0001 should be "morally"—though not mathematically or physically— estimated at zero, thereby yielding zero expectation.

Buffon based his attempts to quantify moral certainty/impossibility on the assumption that "all fear or hope, whose probability equals that which produces the fear of death, in the moral realm may be taken as unity against which all other fears are to be measured." Because no healthy man in the prime of life fears dying in the next 24 hours, Buffon took the probability of such sudden demise, reckoned from the mortality tables to be about .0001, as the zero point ("moral impossibility") on the scale of moral probability. Buffon contended that since no reasonable man gave more than a passing thought to the small but finite probability that he will die tomorrow, he must be equally indifferent to the expectation produced by a .0001 chance of winning the lottery. Indeed, Buffon observed, his chances, while numerically equivalent to his risk of imminent death, hardly balance the psychological intensity of the latter, "since the intensity of the fear of death is a good deal greater than the intensity of any other fear or hope."[18] Here was a psychological standard for expectation, albeit still one linked to reasonable action and belief.

Condorcet's attempts to salvage the classical definition of expectation, which he re-interpreted as an average valid only over the long run, partook of all three types of "moral" considerations: legal, economic, and psychological. In a six-part series of papers published in the *Mémoires de l'Académie royale des Sciences* (1781, 1783, and 1784), Condorcet addressed the foundations and applications of the theory with special attention "to those results too far removed from those given by common reason."[19]

Conceding that mathematical expectation gave absurd results when applied to individual cases, Condorcet concluded that the classical definition held good only for average values. In order to substitute these average values for the "real" values of individual expectation, Condorcet appealed both to jurisprudence and to political economy. According to Condorcet, there existed two possible ways of replacing average values with real ones, both taken from contract law: "voluntary" (a willing exchange of possible for certain gain); and "involuntary" (an unavoidable risk compensated by a certain amount). Condorcet admitted that the two cases differ legally, not mathematically.[20] In both, an uncertain gain is traded for a certain one, or two unequal and unequally probable sums were exchanged. However, the two cases were treated separately because they involved different conditions for equity, although both reduced to the same mathematical conditions.

Voluntary substitutions provided Condorcet with the meat of his analysis. Involuntary substitutions adhered to "the laws of equity," since one need only follow "the sum total of similar conventions, and seek to arrange matters so that the least possible inequality results." In voluntary substitutions, "if one wishes to act with prudence, if the object is important," legal conventions guided actions only insofar as the two parties agreed to a weaker form of equity.[21] Like Bernoulli, Condorcet distinguished between the claims of justice and prudence in defining probabilistic expectation.

Condorcet's discussion of the economic or voluntary substitution of expectations turned upon a comparison between this type of transaction and the exchange of goods in "all other markets." In each exchange of expectations in which commodities have different intrinsic values and therefore could not be rigorously set equal, there must be a personal or subjective "motive for preference" on both sides that impels the trade. Just as reciprocal desires and needs established equality among free agents in the marketplace, so in probabilistic expectation, "neither he who exchanges a certain value for an uncertain one, or reciprocally, nor he who accepts the exchange find in this change any advantage independent of the particular motive of convenience which determined the preference." Averaged over many such exchanges, the expectations, like the common price, would tend toward "the greatest equality possible" between parties.

Condorcet presented the mathematical conditions for a definition of expectation satisfying the rules of the marketplace exchange: over the long run, the most probable outcome should be a net gain or loss of zero for both sides; and as the number of cases becomes very large, the probability of gain or loss on both sides should approach $\frac{1}{2}$. In other words, expectation should be so defined as to give a probability which increases with the number of cases that the advantage of either party over the other will be smaller than any given amount. Condorcet argued that mathematical expectation, conceived as an average value, uniquely satisfied these criteria, to offer "the greatest possible equality between two essentially different conditions."[22] Condorcet therefore dismissed the St. Petersburg problem as an "unreal" case, since the probability of zero net gain or loss for both sides occurs only if the game is repeated an infinite number of times.[23]

A protegé of Condorcet, Laplace was the last and greatest representative of the classical approach to mathematical probability. His writings on expectation reveal that mathematicians had yet to reach a consensus about its definition and application by the turn of the nineteenth century. In a seminal paper of 1774, Laplace attacked the St. Petersburg problem in terms closer to the spirit of d'Alembert's approach, using the analytic formulation of what became known as the Bayes-Laplace theorem on inverse probabilities developed in the same memoir. Laplace accepted d'Alembert's claim that the conventional method of computing expectation in the St. Petersburg problem unrealistically assumed a perfectly fair coin, "a supposition which is only mathematically admissible, because physically there must be some inequality."[24] Supposing that the unknown inequality lay within certain limits, Laplace was able to deploy his method of finding inverse probabilities to compute the "true" expectation. Despite the enthusiasm of d'Alembert and Condorcet for this novel approach, Laplace himself eventually abandoned it in favor of Daniel Bernoulli's moral expectation, at least as far as the St. Petersburg problem was concerned.

Laplace's magisterial *Théorie analytique des probabilités* (1812) systematized all aspects of classical probability theory and expanded many in this definitive formulation of that theory. In the chapter devoted to moral expectation, Laplace adhered closely to Daniel Bernoulli's analysis, citing many of the same examples. Laplace used this method not only for the St. Petersburg problem, but also for problems pertaining to the relative advantages of individual and joint annuities with respect to moral (as opposed to actual) fortune. He concluded this section of the *Théorie analytique* by urging governments to promote such schemes, which fostered "the most gentle tendencies of [human] nature." Furthermore, they were tainted with none of the hidden pitfalls of gambling, which moral expectation exposed as a perpetual losing proposition even in fair games and which sober reflection condemned.

Although Laplace advanced these recommendations on the strength of results derived from moral expectation, the critiques and counter-proposals of d'Alembert and others still colored his views on the subject. While he clung to Bernoulli's hypothesis that the richer one is, the less advantageous a small gain becomes, he warned of the unmanageable complexity of a more comprehensive moral expectation: "But the moral advantage that an expected sum can procure depends on an infinity of circumstances peculiar to each individual, which are impossible to evaluate."[25]

Laplace's reservations concerning the traditional applications of probability theory to problems in the moral sciences were echoed by both social theorists and mathematicians in the first decades of the nineteenth century. His disciple Simon Denis Poisson's attempts to extend the methods of Condorcet and Laplace in his *Recherche sur la probabilité des jugements en matière criminelle et en matière civile* (1837) upheld an approach that had already been criticized by social theorists like the Ideologue Destutt de Tracy and drew further criticisms from mathematicians, including Poisson's colleagues Charles Dupin and Louis Poinsot.[26]

CONCLUSION

The role of mathematical probability in the study of society had changed, as had the moral sciences themselves. The eighteenth-century probabilists, following the precepts of mixed mathematics, had understood the theory as a description and guide to social action based on the example of the archetypal reasonable man. The dictates of "good sense" supplied the data that the calculus of probabilities was to systematize and explain. To the extent that good sense was already codified in jurisprudence or political economy, probabilists found it natural to incorporate concepts and assumptions taken from these disciplines into the mathematical theory. The traffic between mathematical probability and the moral sciences was especially heavy in cases like the St. Petersburg problem, where the mathematical results were at odds with good sense. With the exception of d'Alembert, eighteenth-century mathematicians did not believe that the complexity of the moral realm, as it was reflected in the psychology of decision-making, posed insurmountable obstacles to the probabilist program.

The upheaval of the Revolutionary and Napoleonic era appears to have shaken the confidence of the probabilists in a way that d'Alembert's persistent criticisms and their own disagreements over expectation had not. The conduct of reasonable men no longer seemed an obvious standard, nor a comprehensive basis for a theory of society. Distinguishing prudent from rash behavior in post-Revolutionary France was no easy matter, and just what constituted "good sense" was no longer self-evident. With the demise of the reasonable man, the probabilists had lost both their subject matter and criterion of validity.

By 1840, neither mixed mathematics nor the moral sciences existed in their Enlightenment form. Mathematics was a tool for articulating theory, no longer a description of natural and social phenomena; social theorists studied the properties of whole societies, no longer the conduct of rational individuals.

Although probability theory still featured prominently in the quantitative social sciences, statistical distribution replaced expectation as the central concept in such applications. Whereas eighteenth-century probabilists had sought a mathematical model of the way in which reasonable men made decisions, nineteenth-century statisticians looked for macroscopic regularities in society at-large, often concerning themselves with such unreasonable actions as crime and suicide.

The Enlightenment alliance between mathematical probability and the moral sciences had contained the seeds of its own destruction. I have argued that the eighteenth-century probabilists tailored the mathematical theory to fit the prescriptions of reasonableness. They repeatedly rejected mathematical methods that contradicted the promptings of good sense. "Good sense" was, however, by no means monolithic, admitting several interpretations. Legal, economic, physical, and psychological refinements were all proposed as adjustments to the mathematical theory in the name of good sense. The competing orientations of jurisprudence and political economy created rival definitions of probabilistic expectation, and psychological and physical considerations further splintered mathematical opinion. The controversy was never settled in its own terms; rather, a new brand of social mathematics, Quetelet's statistical "social physics," replaced the old methods and problems altogether. By 1840, probabilists no longer checked their results against the practices and beliefs of "men known for their experience and wisdom in the conduct of their affairs,"[27] and the once neighboring disciplines of jurisprudence, political economy, and mathematical probability drifted ever farther apart.

NOTES AND REFERENCES

1. Large portions of this paper have appeared in my "Probabilistic Expectation and Rationality in Classical Probability Theory," *Historia Mathematica* 7(1980): 234–260.

2. BLAISE PASCAL, *Oeuvres complètes,* J. Mesnard, ed. (Paris: Bibliothèque Européenne, 1970), v. 2, pp. 1132–1158.

3. H. BRUGMANS, *Le séjour de Christian Huygens à Paris et ses rélations avec les milieux scientifiques français* (Paris: André, 1935), p. 40.

4. JAKOB BERNOULLI, *Ars conjectandi* (Basel, 1713), Book I.

5. CHRISTIAN HUYGENS, *Oeuvres complètes* (The Hague: Société Hollandaise des Sciences, 1920), v. 14, pp. 60–61.

6. HUGO GROTIUS, *The Rights of War and Peace,* A. C. Campbell, trans. (Washington and London, 1907), p. 147; ERNST COUMET, "La théorie du hasard est-elle née par hasard?", *Annales: Economies, Sociétés, Civilisations* 25(1970): 574–598.

7. NICHOLAS BERNOULLI, *De usu artis conjectandi in jure* (Basel, 1709), T. Drucker, trans. (unpubl., 1976), p. 26.

8. BERNOULLI,[7] p. 30.

9. *See* THEODORE WALDMAN, "Origins of the Legal Doctrine of Reasonable Doubt," *Journal of the History of Ideas* 20(1959): 299–316, and BARBARA SHAPIRO, "Law and Science in Seventeenth-Century England," *Stanford Law Review* 21(1969): 727–766.

10. ROBERT BOYLE, *Works* (London, 1772), v. 4, p. 182.
11. BOYLE,[10] pp. 184–186.
12. *See* A. O. HIRSCHMAN, *The Passions and the Interests* (Princeton: Princeton University Press, 1977).
13. PIERRE SIMON LAPLACE, *A Philosophical Essay on Probabilities,* F. W. Truscott and F. L. Emory, trans. (New York: Dover, 1951), p. 196.
14. DANIEL BERNOULLI, "Specimen theoriae novae de mensura sortis," *Commentarii Academiae Scientiarum Imperialie Petropolitanae* 5(1730–31; publ. 1738): 175–192; trans. by L. Sommer, "Exposition of a New Theory of Measurement of Risk," *Econometrica* **22**(1954): 23–36.
15. *Ibid.,* p. 24. *See* KENNETH ARROW, "Alternative Approaches to the Theory of Choice in Risk-Taking Situations," *Econometrica* **19**(1951): 404–437.
16. *See* C. B. MACPHERSON, *The Political Theory of Possessive Individualism: Hobbes to Locke* (Oxford: Oxford University Press, 1972), p. 48.
17. JEAN D'ALEMBERT, "Croix ou pile," *Encyclopédie, ou Dictionnaire raisonnée des sciences, des arts et des métiers* (Paris, 1754), v. 4, p. 513.
18. GEORGES LECLERC BUFFON, *Supplément de l'histoire naturelle* (Paris, 1777), v. 4, pp. 56; 58.
19. M. J. A. N. CONDORCET, "Suite de mémoire sur le calcul des probabilités. Article VI," *Mémoires de l'Académie royale des Sciences* (1784, publ. 1787), p. 456.
20. *See,* for example, JEAN DOMAT, *The Civil Law in its Natural Order: Together with the Publick Law,* WILLIAM STRAHAN, trans., 2nd ed. (London, 1737), p. xli.
21. CONDORCET, "Réflexions sur la règle générale qui préscrit de prendre pour la valeur d'un évènement incertain, la probabilité de cet évènement, multipliée par la valeur de l'évènement en lui-même," *Mémoires de l'Académie royale des Sciences* (1784, publ. 1787), pp. 711–712.
22. CONDORCET, "Probabilité," *Dictionnaire encyclopédique des mathématiques* (Paris, 1789), v. 2, pp. 654–655.
23. CONDORCET,[21] pp. 713–718.
24. LAPLACE, *Oeuvres complètes* (Paris, 1891), v. 8, p. 54.
25. LAPLACE,[24] v. 7, p. 449.
26. See *Comptes rendus hébdomadaires des séances de l'Académie des Sciences* 2(1836), pp. 380–381.
27. SILVESTRE FRANÇOIS LACROIX, *Traité élémentaire du calcul des probabilités* (Paris, 1816), p. 257.

Non-Euclidean Geometry and Weierstrassian Mathematics: The Background to Killing's Work on Lie Algebras[a]

THOMAS HAWKINS

Department of Mathematics
Boston University
Boston, Massachusetts 02215

In 1888 a series of papers began to appear in *Mathematische Annalen* entitled "The Composition of Continuous Finite Transformation Groups."[1] The appearance of these papers must have raised some eyebrows because they seemed to constitute a major contribution to mathematics and yet the author, Wilhelm Killing, was a little-known, forty-one-year-old Professor at the Lyceum Hosianum in Braunsberg, East Prussia (now a part of Poland). The Lyceum was a Roman Catholic training center for future clergymen. Despite the unlikely background of the author, the papers did in fact constitute a major contribution to mathematics, a contribution as unexpected as it was extraordinary. It was in these papers that the entire theory of the structure of semisimple Lie algebras originated. Here we find the origins of such key concepts as the rank of an algebra, Cartan subalgebra, Cartan integers, root systems, nilpotent and semisimple algebras, and the radical of an algebra, as well as fundamental results such as the theorem enumerating all possible structures for finite-dimensional simple Lie algebras over the complex field and a radical splitting theorem.

The purpose of my talk is to discuss how such an unlikely figure as Killing came to create such unexpected mathematics. As the title suggests, two factors principally determined the direction of Killing's research. The discoveries in non-Euclidean geometry and the concomitant speculations on the foundations of geometry formed the context of Killing's work. His contributions to the theory of Lie algebras were a by-product of his research program on the foundations of geometry. But when compared with contemporaneous work on non-Euclidean geometry and its foundations, Killing's work stands out as atypical. Since it is precisely the peculiar emphasis of Killing's research program that brings with it the algebraic problem of determining, in effect, all possible structures for Lie algebras, it is of considerable historical interest to seek to understand its basis.

In this connection, Killing's mathematical education is paramount. He

[a]This paper was presented at the December 18, 1979 meeting of the Section of History, Philosophy and Ethical Issues of Science and Technology of The New York Academy of Sciences.

0077–8923/83/0412–0073 $01.75/2 © 1983, NYAS

received his mathematical training in the school of mathematics centered about Karl Weierstrass at the University of Berlin. As will be seen in what follows, during the decade (1867–1877) he spent in Berlin, Killing acquired the mathematical tools and general outlook on mathematics that oriented his geometrical research program in such an unusual manner.

To fully appreciate the magnitude of the impression Berlin made upon Killing it is necessary to go back a bit further in his life. Killing was born and raised in various towns in Westphalia. His aspiration as a *Gymnasium* student was to become a professor of mathematics at such an institution. The prospect of a university professorship apparently appeared too remote to take seriously. To prepare for a teaching career, Killing proceeded to the local university at Münster. The choice of Münster was an unfortunate one because at that time there were no mathematicians on the faculty. Mathematics courses were taught by an observational astronomer who openly admitted that his mathematical training was slight. The courses he gave were consequently elementary. In a sense, they catered to the demands of the students who, much to Killing's disgust, were primarily concerned with learning just enough mathematics to pass the examinations. Killing found himself forced to renew the practice he had found necessary as a *Gymnasium* student: self-study of works by distinguished mathematicians. For example, to supplement the "pablum" presented in the course on analytic geometry, Killing studied the treatises by Plücker and Hesse.

After two years of inferior education, Killing finally left Münster for Berlin. Especially coming from Münster, the impression Berlin made upon Killing must have been spectacular, for the University of Berlin was by then the leading center of mathematics in Germany and possibly in the world. The mathematics faculty, although of modest size by today's standards, consisted of Kummer, Weierstrass, Kronecker, and Fuchs. They were surrounded by a growing number of talented pre- and postdoctoral students. There was also a mathematics union run by students to further the dissemination of mathematical knowledge by means of sponsored lectures, problem-solving contests and the purchase of library books on mathematics. Instead of boring, elementary lectures, Killing could hear challenging lectures by Kummer, Weierstrass, and Kronecker coordinated so as to provide students with a solid background in up to date mathematics. In particular, there was the demanding cycle of lectures on analysis by Weierstrass, with their emphasis upon a rigorous, systematic development of analysis built upon an arithmetical foundation and without recourse to intuition or physical considerations. Finally, there was the famous mathematics seminar, run by Kummer and Weierstrass, in which students were brought to the frontiers of mathematics. (For further details, see the informative study by Biermann.[2])

At Berlin, Killing was attracted above all to Weierstrass, as were many students. But Killing was at heart a geometer whereas Weierstrass was an analyst in the tradition of Lagrange, Cauchy, and Jacobi. Kummer would have been a more logical choice for a doctoral dissertation advisor since 18 out of 31, or 58%, of the dissertations done under his direction were on geometrical topics. Nonetheless, Killing chose to work with Weierstrass. The dilemma of working with him and yet doing something geometrical was resolved by Weierstrass' theory of elementary divisors,[3] which had been published the year following Killing's arrival in Berlin.

The theory of elementary divisors is concerned with necessary and sufficient conditions that one family of bilinear forms $A - \lambda B$ ($|B| \neq 0$) be transformable by means of linear variable changes into another such family. The theory is thus equivalent to the more familiar theory of canonical matrix forms, and in fact the elementary divisors of the characteristic polynomial $|A - \lambda B|$ are the divisors corresponding to the Jordan blocks of the Jordan canonical form for $A - \lambda B$. Weierstrass actually introduced this canonical form in his paper on elementary divisors and therefore prior to Jordan.[4]

The theory of elementary divisors represented Weierstrass' critical response to the prevalent tendency in eighteenth and nineteenth century algebraic analysis to reason vaguely in terms of a sort of "general case," according to which the algebraic symbols involved are regarded as assuming "general" rather than specific values.[4-6] This mode of reasoning brought with it a tendency to overlook, or ignore, potential limitations to the conclusions of a mathematical argument. This tendency was widespread and not limited to inferior mathematicians. Lagrange, Laplace, Jacobi, Hermite, and Cayley were among the practitioners of "generic" analysis. Within the context of the transformation of quadratic and bilinear forms, where "in general" the roots of characteristic polynomials are all distinct, Weierstrass rejected the tenability of such an attitude. An exhaustive analysis of all the algebraic possibilities was demanded by Weierstrass and facilitated by his theory of elementary divisors.

In Berlin the theory of elementary divisors was thus regarded as proving more than mathematical theorems. It demonstrated the desirability and feasibility of a more rigorous approach to algebraic analysis, one that did not shy away from the multitude of special cases that present themselves when the "general case" is abandoned. The following passage from a paper of 1874 by Kronecker[7] exemplifies this attitude towards Weierstrass' theory. After referring to it, Kronecker wrote:

> It is common, especially in algebraic questions, to encounter essentially new difficulties when one breaks away from those cases customarily designated as general. As soon as one penetrates beneath the surface of the so-called generality that excludes every particularity into the true generality, which includes all singularities, the real difficulties of the investigation are usually first encountered but, at the same time, also the wealth of new viewpoints and phenomena which are contained in its depths. (p. 405)

It was in this spirit that Killing proposed to use Weierstrass' theory of elementary divisors to study, as his doctoral dissertation, pencils of quadric surfaces.[8] In homogeneous coordinates, such a pencil corresponds to a family of quadratic forms $A - \lambda B$ in four variables. The various geometrical possibilities can be systematically explored by first of all classifying according to the thirteen possibilities for the elementary divisors of the characteristic polynomial, and this is the approach taken by Killing. In the preface to his dissertation Killing explained that his intention was simply to provide a *geometrical interpretation* of his mentor's theory of elementary divisors. But, in effect, Killing also gave a geometrical interpretation to the Weierstrassian approach to mathematics: In geometrical investigations also one must systematically explore all the possibilities revealed by the analytical framework as viewed from the Weierstrassian standpoint rather than in "general" terms. This was not a widespread attitude among geometers of the period. As will be seen, Killing attacked the foundations of geometry in the same spirit as quadric surfaces.

Killing received his doctorate in March of 1872, but remained in Berlin until 1878, teaching at the *Gymnasium* level. During the summer semester 1872 the new doctor of mathematics participated in the mathematics seminar and it was there that he heard Weierstrass present some lectures on the subject that was to become his life work: the foundations of geometry. Weierstrass was, of course, interested in foundational matters, but this subject was also a timely one because several events had combined to focus the interest of mathematicians and philosophers on non-Euclidean geometry and related foundational issues. The publication, during the mid-1860s, of Gauss' extensive correspondence with the astronomer Schumacher served to rescue the work of Lobachevsky from oblivion; for although Gauss carefully refrained from publicly expressing his favorable estimate of Lobachevsky's geometry while alive, his letters to Schumacher conveyed his high opinion of Lobachevsky's work. Then in 1868 Beltrami showed that plane Lobachevskian geometry could be interpreted on a surface of constant negative curvature, thereby convincing sympathetic mathematicians that Lobachevsky's geometry did not contain hidden contradictions. Also in 1868 the philosophically oriented essays of Riemann[9] and Helmholtz[10] on the foundations of geometry appeared.

The Berlin Seminar of 1872 opened the door to a new mathematical world for Killing, the realm of non-Euclidean geometry. It will be helpful at this point to indicate the principal discoveries in this realm and their significance from Killing's perspective. The discovery of Lobachevskian geometry showed that the geometry of Euclid was not the only geometry both logically consistent and, in so far as could be determined, compatible with experience. The discoverers of Lobachevskian geometry (e.g. Lobachevsky, Bolyai, and Gauss) tended to regard it in absolute times as exhausting the possibilities for geometry. That is, Lobachevskian geometry contained a parameter k and for $k = \infty$ Euclidean geometry is obtained. Thus Lobachevskian geometry, including therewith the limiting case, embraced Euclidean geometry, and, in the eyes of its discoverers, all conceivable geometrical possibilities.

When Riemann, in his celebrated essay, suggested the possibility of a geometry of "finite space" corresponding to a manifold of constant positive curvature because the unboundedness of space was more certain that its infinitude, he therefore caused something of a sensation. Although, spherical geometry had long been familiar, Riemann was apparently the first to seriously consider this sort of geometry as an alternative to the geometry of Euclid. Whether Riemann actually intended to identify his geometry of finite space with spherical geometry is not certain due to the vagueness of his primarily non-mathematical description, but most mathematicians, including Beltrami and Weierstrass, assumed Riemann was speaking of spherical geometry. Given this interpretation, Riemann's geometry of finite space was observed to violate the "axiom of the straight line" that two points *uniquely* determine a straight line. No sooner had the violation been assumed a necessary characteristic of a geometry of finite space (or of constant positive curvature) than Felix Klein and Simon Newcomb pointed out, independently, that there exists another geometry of finite space in which the axiom of the straight line does hold. Klein called it elliptical geometry.

Each new discovery thus revealed the limitations of the previous conceptions of the geometrical possibilities, a point that made a great impression upon Killing. The situation in geometry was analogous to what was occurring in the foundations of analysis, where earlier conceptions and intuitions regarding, for example, the properties of continuous functions were gradually being undermined by discoveries such as that of Weierstrass' example of a continuous nowhere differentiable function.[11] Since Weierstrass presented his example to the Berlin Academy in July of 1872, it is likely that he also presented it in the same mathematics seminar where Killing had heard him lecture on the foundations of geometry. The analogy between developments in geometry and the theory of functions tended to reinforce Killing's attitude towards the foundations of geometry. In fact, Killing invoked the existence of functions such as Weierstrass' example to refute the arguments of one geometer who attempted to demonstrate that there existed exactly three possible geometries: Euclidean, Lobachevskian, and spherical.

The discoveries in non-Euclidian geometry, reinforced by the discoveries in real analysis, convinced Killing that the science of geometry must be conceived in very general terms. For example, Killing concluded that it was impossible to completely capture spatial intuitions in an axiom system since diverse "models" (as we would say) can represent the same system. Euclid's system can be realized on the surface of zero curvature or by Klein's projective model (parabolic geometry). Nor could an axiom system uniquely capture the human experience of space. According to Killing, there were (at least) four geometrical systems compatible with experience: Euclidean, Lobachevskian, spherical, and elliptical.

Killing therefore concluded that the science of geometry is necessarily very general, its only requirement being logical consistency and completeness in the sense of embracing all the geometrical possibilities. Consequently:

> We regard all investigations of this type as branches of the same science and designate every individual possibility, with its further consequences, as a space form. Many may directly contradict experience, others may be very unlikely, but they nonetheless all rest on the same foundations and exhibit in their proof procedures an unmistakable similarity.[12] (p. 4)

In the spirit of Weierstrass, Killing proposed an exhaustive analysis of all the geometrical possibilities—all possible space forms—without recourse to the delimiting deceptions of intuition and experience, including thereby geometries that could be as counter-intuitive as continuous, nowhere-differentiable functions.

Among geometers, Killing's attitude was not at all typical. No one, including Killing, took seriously Riemann's vague hints that physical science might require a geometry corresponding to a manifold of variable curvature. Experience, the seemingly essential homogeneity of space, required a geometry corresponding to a manifold of constant curvature. Thus Riemann's greatest contribution to geometry was, in the view of Felix Klein, the concept of a manifold of *constant* curvature. Such a manifold formed the analytical context for geometry and limited the possibilities to the known types of Euclidean and non-Euclidean geometries. Because the distinction between the local and global characteristics of a geometry was glossed over for a long time, the geometrical possibilities seemed limited to most geometers, although at the same time it must be remembered that non-Euclidean geometry was still (in 1860–1880) a radically new subject, a new mathematical world awaiting exploration and in this sense did not give the impression of being "limited."

Nonetheless, Killing's conception of the scope of geometry, of the geometrical possibilities, was considerably more far-reaching than that of his contemporaries. Only Riemann, with his idea of a manifold of variable curvature, entertained a view of geometry of comparable generality. But Killing's motivation was entirely different. Killing did not wish to transcend the limits of geometry as conceived by his contemporaries in the belief (such as Riemann held) that a more general conception of geometry might prove necessary to deal with physical reality. In arguments, Killing admitted that Euclidean geometry was the "only true" geometry, and he always regarded the well-known geometries of constant curvature as the most important because they corresponded to experience. But these sentiments did not diminish the need Killing felt to transcend experience and intuition in order to secure the foundations of geometry. This was of course precisely Weierstrass' attitude towards the foundations of analysis.

In their essays on the foundations of geometry both Riemann and Helmholtz made assumptions about the metrical properties of space. Riemann had argued tentatively that metrical relations were determined by a quadratic differential form; and Helmholtz, who approached geometry through the motions of rigid bodies, attempted a demonstration that his axioms on mobile rigid bodies, which included the assumption of some sort of a distance function left invariant by motions, implied that infinitesimal distances are expressible as a quadratic differential form as Riemann claimed. The treatment of these matters by both Riemann and Helmholtz was, however, lacking in clarity and rigor. It was probably because of this that Weierstrass, in his seminar lectures in 1872, called for a further exploration of possible metrics and the resulting geometries.

Killing went a step further and decided it would be best to begin without any assumptions of a metrical nature:

> Attempts to create a natural foundation for geometry have hitherto not been accompanied by the desired success. The reason for this lies, in my opinion, in this: just as geometry had to abandon the concept of direction in the sense stipulated by the parallel axiom, so also the concept of distance cannot be maintained as a general basic concept and therewith [geometry] must go far beyond the non-Euclidean space forms in the narrower sense [of Euclidean, Lobachevskian, spherical and elliptical geometry].[13] (p. iv)

Following Helmholtz, who incidentally was Professor of Physics at Berlin from 1870 onward, Killing approached the foundations of geometry through the motions of rigid bodies but without introducing metrical concepts. Killing's objective was to explore systematically and analytically the possibilities for his general space forms and, once this was achieved, then to consider the question of metrical properties.

The analytical starting point of Killing's theory of general space forms is the notion of an n-dimensional "manifold" of points $x = (x_1, \ldots x_n)$ endowed with "m degrees of mobility." In the spirit of Riemann and Helmholtz, Killing concentrated on the infinitesimal aspects of the geometry and thus he dealt exclusively with infinitesimal notions $x \rightarrow x + dx$, where

$$dx_\rho = u^{(\rho)} \, dt, \, u^{(\rho)} = u^{(\rho)}(x).$$

Certain properties of these motions followed from generally accepted practices. Thus if $dx = udt$ is a motion, so is $dx = \lambda udt$. According to Killing, it is essentially the same motion if velocity is ignored. Likewise if $dx = udt$ and $dx = vdt$ are motions, their "composite" $dx = (u + v) \, dt$ is another motion. From the traditional point of view, the composition of infinitesimal motions is commutative (as we would say).

The generally accepted properties of infinitesimal motions were thus used by Killing to assert, in modern language, that the infinitesimal motions of a space form constitute a vector space. Killing assumed that this vector space had a finite dimension, m, which he called the degree of mobility of the space form. Killing's objective was thus to determine all possible space forms in n-dimensions with m degrees of mobility.

In the commonly accepted Euclidean and non-Euclidean geometries, one had $m = n(n + 1)/2$, but Killing of course, imposed no such restrictive condition on his space forms. The problem of exhaustively determining the possibilities was thus a difficult one and although Killing did not really appreciate how difficult it would prove to be, he did sense the need to go beyond the traditional treatment of infinitesimal motions to make his problem tractable. In this connection, he observed that although, traditionally, the *order* of the composition of infinitesimal motions does not matter—everything commutes—this is not true of the "finite" motions of a geometry. He thus sought to capture this "non-commutativity" at the infinitesimal level. The manner in which he did this is obscure in an interesting way because Killing needed to invoke the traditional views on infinitesimal motions while, in effect, denying their adequacy.

As was customary in the nineteenth century, Killing reasoned in terms of a basis, in this case a basis $dx = u_i dt$, $i = 1, 2, \ldots, m$, for the infinitesimal motions of the space form. Given two such motions, $dx = u_i dt$ and $dx = u_k dt$, according to the traditional interpretation, a point x is moved to $x + dx$ by the resultant composite motion, where $dx = (u_i + u_k)dt$. Although Killing did not mention this explicitly, he probably observed that if M_i and M_k denote the corresponding finite motions, it can happen that $M_i M_k x \neq M_k M_i x$ and thus $(M_k M_i)(M_i M_k)^{-1}$ does not leave x fixed. Presumably this is why Killing decided that, at the infinitesimal level, a nontrivial motion rather than the trivial $dx = 0$ implied by the traditional interpretation should correspond to the commutator motion $(M_kM_i)(M_iM_k)^{-1}$. We now consider how Killing obtained an infinitesimal analog of the commutator.

Consider first the infinitesimal analog of M_kM_i. The motion $dx = u_i d\sigma$ sends x into $x + dx$, where, coordinatewise:

$$y_\rho = x_\rho + u_i^{(\rho)}(x)d\sigma, \qquad \rho = 1, \ldots, n. \tag{1}$$

The motion $dx = u_k dt$ then sends y into $z = y + dy$, where $dy = u_k^{(\rho)}(y)dt$. If, as was customary, $u_k^{(\rho)}(y) = u_k^{(\rho)}(x + u_i d\sigma)$ is expanded in a Taylor series and higher order terms neglected, the result is:

$$dy_\rho = (u_k^{(\rho)}(x) + \sum_{\nu=1}^{n} \frac{\partial u_k^{(\rho)}}{\partial x_\nu} u_i^{(\nu)}d\sigma)dt \tag{2}$$

Thus the coordinates of the point $z = y + dy$ are:

$$z_\rho = x_\rho + u_i^{(\rho)}(x)d\sigma + u_k^{(\rho)}(x)dt + d\sigma \left(\sum_{\nu=1}^{n} \frac{\partial u_k^{(\rho)}}{\partial x_\nu} u_i^{(\nu)} \right)dt. \tag{3}$$

From the traditional standpoint, of course, the last term should be neglected since it is a second order infinitesimal. The resultant equation (3) would then yield the usual interpretation of composition. Killing did not drop the term, although by separating $d\sigma$ and dt as he did, he seemed to wish to minimize the visibility of his unconventional procedure.

The infinitesimal analog of M_iM_kx is obtained by the same procedure. Thus $x \rightarrow w$, where

$$w_\rho = x_\rho + u_i^{(\rho)}(x)d\sigma + u_k^{(\rho)}(x)dt + d\sigma \left(\sum_{\nu=1}^{n} \frac{\partial u_i^{(\rho)}}{\partial x_\nu} u_k^{(\nu)} \right)dt. \tag{4}$$

Now *by the traditional convention*, the inverse of $x \rightarrow x + dx$ is $x \rightarrow x - dx$. Accepting this convention, Killing obtained the infinitesimal motion $w \rightarrow z$ corresponding to $(M_kM_i)(M_iM_k)^{-1}w$ by subtracting (4) from (3) to obtain (with a convenient omission of the telltale dt):

$$dx_\rho = U_{ik}^{(\rho)}d\sigma, \qquad U_{ik}^{(\rho)} = \sum_{\nu=1}^{n} \left(u_i^{(\nu)} \frac{\partial u_k^{(\rho)}}{\partial x_\nu} - u_k^{(\nu)} \frac{\partial u_i^{(\rho)}}{\partial x_\nu} \right). \tag{5}$$

Since the motions $dx = u_i dt$, $i = 1, \ldots, m$, form a basis for all motions of the space form, this new motion (5) must be expressible in terms of them. That is, constants $a_{j,ik}$ must exist so that

$$U_{ik} = \sum_{j=1}^{m} a_{j,ik}u_j \ldots \tag{6}$$

A total of m^2 motions $dx = U_{ik}d\sigma$ are produced in this manner, but at most m of them can be linearly independent. Realizing this, Killing sought relations among the U_{ik}. He singled out the following:

$$U_{ik} + U_{ki} = 0, \qquad U_{ii} = 0 \tag{7}$$

$$\sum_{j=1}^{m} (a_{j,kl}U_{ij} + a_{j,li}U_{kj} + a_{j,lk}U_{lj}) = 0. \tag{8}$$

Equations 6–8 look more familiar when translated into differential operator notation. Associate with the motions u_i and u_k the respective operators

$$X_i = \sum_{p=1}^{n} u_i^{(p)} \frac{\partial}{\partial x_p} \quad \text{and} \quad X_k = \sum_{p=1}^{n} u_k^{(p)} \frac{\partial}{\partial x_p}.$$

Then $[X_i X_k]$ corresponds to the motion U_{ik} and (6) corresponds to

$$[X_i\, X_k] = \sum_{j=1}^{m} a_{j,ik}\, X_j,$$

while (7) and (8) correspond respectively to:

$$[X_i\, X_k] + [X_k\, X_i] = 0, \qquad [X_i\, X_i] = 0$$

$$[X_i[X_k\, X_l]] + [X_k[X_l\, X_i]] + [X_l[X_i\, X_k]] = 0.$$

The infinitesimal motions of a space form thus constitute, in effect, a finite dimensional Lie algebra.

The problem of determining all space forms therefore involved three substantial subproblems. First there was the algebraic problem of determining all nonequivalent possibilities for the "structure constants" $a_{j,ik}$ or, in other words, the problem of determining all nonisomorphic Lie algebras. In the algebraic problem the u_i are treated as symbols, not specific functions of x. The analytic problem is to determine these functions by solving the partial differential equations given by equations 5 and 6. The third and final problem was geometrical: given the functions $u_i = u_i(x)$ and thus the infinitesimal motions of the space form, describe its geometrical characteristics.

The problem of providing an exhaustive determination of space forms was thus analogous to, albeit infinitely more difficult than, the problem resolved by Killing in his doctoral dissertation.[8] In the present problem as well, the theory of elementary divisors enters in a natural way. In order to integrate the partial differential equations (5) and (6) it is desirable to choose constants $a_{j,ik}$ so that most are zero. One way to accomplish this is to begin by choosing a basis as follows. Choose, using the operator notation, X_1 arbitrarily (at least provisionally). Then choose the remaining X_i so that the linear transformation $X \to [X_1 X]$ is in its Weierstrass-Jordan canonical form. By this route Killing commenced to develop what was to become the theory of the structure of Lie algebras and especially semisimple Lie algebras.

Killing's theory of space forms as just described was published in 1884 as a *Programmschrift*,[14] i.e. a scholarly essay appended, as was the custom, to the schedule of courses to be offered at the Lyceum during the forthcoming semester. It was consequently not widely read, if read at all. But Killing fortunately sent a copy to Klein, who, being a friend of Sophus Lie, suggested a possible connection between Killing's space forms and Lie's theory of continuous transformation groups, a theory he had been developing since 1874, although it was still (in 1884) not well known in Germany since Lie published most of his work in an inaccessible Norwegian journal that he edited. Killing finally obtained copies of these publications on loan from Lie for a relatively short time and he never really digested their contents. He read through them primarily to see where he had been anticipated by Lie, and he discovered that his own methods were not employed by Lie. Through his contact with Lie and Lie's assistant, Friedrich Engel, however, Killing's research received a needed focus. In 1885 Lie published a paper in *Mathematische Annalen*[15] on the application of transformation groups to differential equations, and stressed the importance, in the Galois type theory he envisioned, of being able to specify all simple groups. Lie's methods were not suited to achieving such a specification, and Killing felt that his own methods might succeed here. And of course as we all know he was correct!

Postscript. The above presentation is based upon a more extensive and carefully documented study,[16] which, in particular, deals with a matter passed over here in silence: the apparent similarities between Killing's theory of space forms and Klein's *Erlanger Programm* of 1872. A careful examination of Klein's attitude towards mathematics and its relation to science and intuition suggests that the similarities are superficial and belie a more fundamental lack of affinity between Klein's *Erlanger Programm* and Killing's *Braunsberger Programm*.[14] Although Klein did characterize the study of geometry as the study of transformation groups acting on manifolds, he did so in the hope of bringing unity to the seemingly disparate geometrical theories then proliferating and with no intention of calling for an exhaustive determination of all such groups in order to establish the foundations of non-Euclidean geometry. Geometry, and mathematics in general, was never to be pursued on a level or in a manner incompatible with intuition, experience and the needs for science. As the following passage from Klein's lectures on non-Euclidean geometry[17] illustrates, he was totally unsympathetic to the attitude towards mathematics fostered by Weierstrass and his colleagues at Berlin:

> With what should the mathematician concern himself? Some say: certainly intuition is of no value whatsoever; I therefore restrict myself to the pure forms generated within myself, unhampered by reality. That is the password in some places in Berlin. By contrast, in Göttingen the connection of pure mathematics with spatial intuition and applied problems has always been maintained and the true foundations of mathematical research recognized in a suitable union of theory and practice. (Vol. II, p. 361)

Despite the exaggeration, Klein's words captured the spirit in which mathematics was pursued in Berlin. The constrasts he drew between Berlin and Göttingen apply in particular to Killing's and his own approach to geometry.

REFERENCES

1. KILLING, W. 1888–1890. Die Zusammensetzung der stetigen endlichen Transformationsgruppen. Math. Ann. **31:** 252–290; **33:** 1–48; **34:** 57–122; **36:** 161–189.

2. BIERMANN, K-R. 1973. Die Mathematik und ihrer Dozenten an der Berliner Universität 1810–1920. Akademie-Verlag. Berlin, GDR.

3. WEIERSTRASS, K. 1868. Zur Theorie der quadratischen und bilinearen Formen. Monatsber. Akad. Wiss. Berlin.:311–338.

4. HAWKINS, T. 1977. Weierstrass and the theory of matrices. Archive for History of Exact Sciences **17**: 119–163.

5. HAWKINS, T. 1975. Cauchy and the spectral theory of matrices. Historia Mathematica **2**: 1–29.

6. HAWKINS, T. 1977. Another look at Cayley and the theory of matrices. Archives Internationales d'Histoire des Sciences **26**: 82–112.

7. KRONECKER, L. 1874. Über Schaaren von quadratischen und bilinearen Formen. Monatsber. Akad. Wiss. Berlin.:349–413.

8. KILLING, W. 1872. Der Flächenbüschel zweiter Ordnung. Inaugural-Dissertation. Berlin.

9. RIEMAN, B. 1868. Über die Hypothesen, welche der Geometrie zu Grunde liegen. Abhandlungen K. Ges. Wiss. Göttingen **13**.

10. HELMHOLTZ, H. 1868. Über die Thatsachen, die der Geometrie zum Grunde liegen. Nachrichten K. Ges. Wiss. Göttingen, Nr. 9.

11. HAWKINS, T. 1975. Lebesgue's Theory of Integration: Its Origins and Development. 2nd edit.: 42–54. Chelsea Pub. Co. New York.

12. KILLING, W. 1880. Grundbegriffe und Grundsätze der Geometrie. Programm des Gymnasiums zu Brilon. Brilon.

13. KILLING, W. 1885. Die nichteuklidischen Raumformen in analytischer Behandlung. Leipzig.

14. KILLING W. 1884. Erweiterung des Raumbegriffes. Programm Lyceum Hosianum. Braunsberg.

15. LIE, S. 1885. Allgemeine Untersuchungen über Differential gleichungen, die eine kontinuierliche, endliche Gruppe gestatten. Math. Ann. **25**: 71–151.

16. HAWKINS, T. 1980. Non-Euclidean Geometry and Weierstrassian Mathematics: The Background to Killing's Work on Lie Algebras. Historia Mathematica **7**: 289–342.

17. KLEIN, F. 1893. Nicht-Euklidische Geometrie. Vorlesung, gehalten während 1889–1890. Ausgearbeitet von Fr. Schilling. Göttingen.

Bell Laboratories: The Beginnings of Scientific Research in an Industrial Setting[a]

DEIRDRE LAPORTE

Bell Laboratories, Room WB 2A-111
Holmdel, New Jersey 07733

The establishment of an archives at Bell Laboratories seem to be a fitting time to reflect upon the traditional accounts of the company's origins and achievements. It is to these accounts that the archives ultimately owes its existence, and it is these accounts that will be tested by future research with archival materials.

The Bell System has long been conscious of its leading role in the development of telecommunications technology, of the important contributions by its staff to scientific research, and of the impact both of these have had on society. One result of this awareness has been a stream of historical chronologies and narratives both oral and written. The first two sketches that follow—of the evolution of the corporate structure and of the institutionalization of discovery and invention in the Bell System—are based upon these traditional accounts. They are then supplemented by brief discussions of the archival holdings of Bell Laboratories, of some difficulties to be faced in setting up an archives in an industrial research laboratory, and of the special advantages enjoyed by Bell Laboratories.

EVOLUTION OF THE CORPORATE STRUCTURE: FROM "THE BELL PATENT ASSOCIATION" TO AT&T[b]

On 7 March 1876, Alexander Graham Bell, a teacher of the deaf and a proponent of "Vocal Physiology and the Mechanics of Speech," was issued U.S. Patent No. 174,465 for "An Improvement in Telegraphy." Together with an earlier patent (issued 6 April 1875) and two later ones (issued 6 June 1876 and 30 January 1877), this document formed the legal technological basis of what became the Bell System.

For some years, much of Bell's spare time had been devoted to telegraphy, an interest he shared with many contemporaries; specifically, he was trying to improve telegraphic communication by finding a way to send several coded

---

[a]This paper was presented at the January 28, 1981 meeting of the Section of History, Philosophy and Ethical Issues of Science and Technology of The New York Academy of Sciences.

[b]This narrative of the corporate history of the Bell System is mainly based on Langdon (1923) and Fagen (1975) as well as other Bell System publications included in the list of references.

0077–8923/83/0412–0085 $01.75/2 © 1983, NYAS

messages simultaneously over one wire. Since the fall of 1874, his work on the multiple or harmonic telegraph had been supported financially by two wealthy patrons in return for a share of whatever patents might result. Bell's first patron, Thomas Sanders of Salem, Massachusetts, was a leather merchant and the father of one of Bell's deaf pupils. Bell's second patron, Gardiner Greene Hubbard, an attorney of Cambridge, later became his father-in-law. On 27 February 1875, Bell, Sanders, and Hubbard had signed a written agreement, forming what has been called "The Bell Patent Association" (Langdon 1923, pp. 134–135).

By the summer of 1877, however, another of Bell's long-standing interests, the electrical transmission of speech, seemed to have enough promise for these three men to organize a joint stock company, the Bell Telephone Company, to manufacture telephones and to license their use under the four Bell patents. (Hubbard borrowed the idea of licensing equipment rather than selling it from the Gordon McKay Shoe Machinery Company, which he served as attorney [Langdon 1923, p. 139].)

On 12 February 1878, Sanders and a group of men from Massachusetts and Rhode Island incorporated the first associated telephone company, "to carry on the business of manufacturing and renting telephones and constructing lines of telegraph therefor, in the New England States" (Langdon 1923, p. 141). As trustee of the Bell Telephone Company, Hubbard assigned to this company rights under the four Bell patents for use within New England. In return, the New England Telephone Company (not the modern company of that name) agreed to buy its telephones from the Bell Telephone Company, to lease the instruments to its subscribers, and "to cooperate in the establishing of connecting lines and in the joint working of the same" (Langdon 1923, p. 142).

Later in 1878, the Bell Telephone Company was refinanced and reorganized with G. G. Hubbard as president, Thomas Sanders as treasurer, Charles E. Hubbard (brother of Gardiner G.) as clerk, and Theodore Newton Vail (formerly head of the U.S. Railway Mail Service) as general manager. This was the first of a series of reorganizations that marked the early years of the Bell System (see FIG. 1). With the reorganization, capitalization of $450,000 was acquired by issuing 4500 shares of stock with a par value of $100 each.

But before the end of 1878, more money was needed. William H. Forbes, a railroad man and (like Gardiner Hubbard) antimonopolist, was brought in as a director (Solomon 1978, pp. 147–148). In early 1879, New England Telephone and Bell Telephone were merged to form the National Bell Telephone Company, with headquarters in Boston, Forbes as president, and a capitalization of $850,000.

This consistent need for more and more capital testified to the rapid growth of the telephone business. As the telephone spread rapidly—if haphazardly—across the United States, local corporations were formed and entered into licensing agreements with National Bell to lease instruments.

Growth was one sign of the success of the telephone—others were competition and litigation. In 1877, the wealthy and powerful Western Union Telegraph Company had refused an offer to buy the Bell patents for $100,000. Less than a year later, however, Western Union aggressively entered the telephone business, having acquired the rights to Elisha Gray's "talking telegraph" and Thomas Alva Edison's superior carbon transmitter. The challenge to the Bell company became even more serious when, during the summer of 1879, Western Union bought control of various local companies that were operating with licenses from National Bell.

Under the vigorous leadership of General Manager Vail, Bell Telephone and its successor National Bell fought back. The weapons were both technological—an improved transmitter—and legal—a suit against Western Union charging patent infringement.

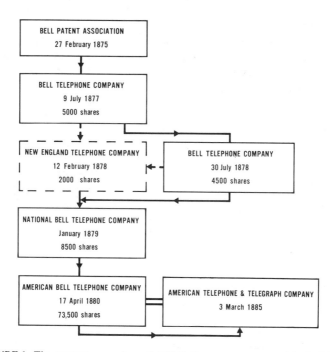

FIGURE 1. The corporate genealogy of AT&T, the parent company of the Bell System.

In November 1879, an out-of-court settlement was reached. Western Union acknowledged the validity of the Bell patents. The telephone business, including the patents held by Western Union, was assigned to National Bell, and the telegraph business, plus royalties, to Western Union.

As a consequence of this settlement, once again the need for coordination of service, reorganization, and refinancing was apparent. In March 1880, by a special act of the Massachusetts legislature, the American Bell Telephone Company was created, with Forbes as president, Vail as general manager, and a capitalization not to exceed $10 million.

While National Bell had succeeded in coordinating the lines of the rapidly growing number of telephone subscribers on the local level by means of the exchange system, the need for long-distance lines to interconnect these local exchanges was becoming more pressing. The first long-distance line, between Boston and New York, was completed in 1884. Soon it was obvious that the construction of a national system of telephone service would require a good deal more money than was available under Massachusetts law. Therefore, in February 1885, the American Telephone and Telegraph Company was chartered in New York State specifically to construct lines throughout the North American continent and to make connections with the rest of the world. For the first 15 years of its existence, AT&T was a subsidiary of American Bell and was known as the "long-distance company." Because of continuing financial restrictions in Massachusetts, American Bell transferred all of its assets to AT&T and by 1900 effectively passed out of existence.

In 1887, Theodore Vail resigned from the Bell System because of disagreements with Forbes about whether profits ought to be used for expanding service (Vail) or be distributed as dividends (Forbes). Then in 1893 and 1894, the Bell patents expired, ending the Bell System's telephone monopoly. Competitors were now free to build duplicate exchanges or to extend service into areas—largely nonurban—not reached by Bell.

By 1907, AT&T owned only about half of the telephones in service, and the company's profitability had been drastically reduced. During the Panic of 1907, banking interests, headed by J. P. Morgan, gained control. They immediately rehired Theodore Vail, who took control on 1 May 1907.

Vail set out to reorganize the company's finances, its general management, and its engineering and research. To remain competitive, Bell had to improve service, reduce rates, and buy up competing exchanges. The Vail era, from 1907 to 1918, put an end to financial instability and shaped the Bell System into its familiar form according to Vail's ideal of "One Policy, One System, and Universal Service" (Langdon, 1923, p. 152).

THE INSTITUTIONALIZATION OF DISCOVERY AND INVENTION IN THE BELL SYSTEM[c]

The initial concept on which the Bell telephone patents were based was Alexander Graham Bell's own—as was his confidence that the telephone was not a toy, that it, as well as the multiple telegraph, was worth working on. Bell's work in telegraphy was a result of his fascination, as a student of the physiology of speech and hearing and a professional teacher of the deaf, with devices that might transmit or reproduce sound (e.g., the phonautograph at MIT). These interests in actually teaching the deaf to speak and in the mechanics of speech and hearing eventually led Bell to consider seriously the possibility of reproducing and transmitting, electrically, not just sounds and codes but actual speech.

While the concept was his own, Bell shared the experimental work. Like other enthusiasts of the new communications technology, he frequented the

---

[c]This account of the origins of research and development in the Bell System is largely based on Bell System sources, including Fagen (1975) and AT&T (1958/1979). I am also indebted to Reich (1977) and Hoddeson (1981).

Boston shop of Charles Williams, Jr., a maker of telegraph instruments. Williams assigned a machinist, Thomas A. Watson, to help Bell. By early June 1875, Bell and Watson had succeeded in transmitting from one room to another sounds that while unintelligible were recognizable as speech. Experiments continued through that summer to improve the "electrical speaking telephone." On 10 March 1876, three days after Bell's second patent was issued, while Bell and Watson were working with a variable-resistance transmitter,[d] an audible sentence, the first telephone message, was sent: "Mr. Watson, come here. I want you."

Bell was not a professional inventor, nor was he just a "telephone man." Once the telephone patents had been obtained, a functioning instrument built, and a company set up to exploit the patents, he effectively retired from the telephone business to pursue other, broader interests.

On 1 September 1876, Hubbard, Sanders, and Bell gave Watson a contract and a one-tenth share of the profits to oversee the manufacture of telephones, to develop subsidiary apparatus, and to solve whatever problems might arise as the very first telephone connections were made. For four years, until June 1881, Watson, as general inspector, had complete charge of all the technical work provided to the Bell Patent Association and its successors: the Bell, National Bell, and American Bell Telephone Companies (Langdon 1923, p. 138).

At first, the research and development that were needed were of the most practical, and yet the most far-reaching, sort. They included the fundamentals of instrument design and testing, installation, and interconnection, that is, the creation from scratch of the basic mechanisms for signaling, transmission, reception, and switching—the first telephone *system*. (The only model available—and it was far from satisfactory—was the telegraphic system.)

Some of this early technical work made the telephone more convenient to use: for example, the separation of the transmitter and receiver, so subscribers did not have to keep moving one rather bulky box from mouth to ear and back again, and the invention of a call signal. Some work ensured the survival of the Bell Telephone Company itself. In 1878, for example, when the company faced the serious challenge mounted by Western Union, the Bell strategy included finding a transmitter equal to or better than Thomas Edison's carbon-button transmitter. Bell purchased the rights to loose-contact transmitters invented by Emile Berliner and Francis Blake and set about adapting them for commercial use. The improved Blake transmitter became the Bell System standard (Fagen 1975, pp. 69–70).

At first, telephones were manufactured and additional equipment designed in Charles Williams, Jr.'s shop, under Watson's direct supervision. In 1878, Watson was joined by Emile Berliner. As chief inspector of the telephone and transmitter testing division, Berliner had two assistants (J. H.

---

[d]The variable-resistance, or liquid, transmitter was combined with a tuned-reed receiver and a battery. The diaphragm of this transmitter was attached to a wire, the other end of which was inserted into a metal cup full of acidulated water. Speaking into the horn of the transmitter made the diaphragm vibrate, which moved the wire up and down in the acid solution, thereby varying the resistance of the circuit.

Cheever and W. L. Richards) and worked on improving his own and Blake's transmitters (Fagen 1975, p. 30; AT&T 1958/1979, p. 6). In 1879, George L. Anders and Thomas Lockwood joined Watson's staff (AT&T 1958/1979, p. 8). Anders specialized in bells and other signaling devices. Lockwood's duties as assistant general inspector included opening exchanges in New England, New York, and New Jersey and visiting other exchanges to offer technical assistance (Lockwood 1926, p. 495). In 1880, W. W. Jacques, who had a Ph.D. from Johns Hopkins and had studied abroad in Germany and Austria, joined the technical staff as an experimenter (Hoddeson 1981, p. 519). Jacques was the first academically trained scientist at Bell; his predecessors and colleagues were all inventors and mechanics.

By 1879, the demand for new instruments was such that licenses to manufacture telephone equipment under the supervision of Watson and his associates were issued to firms such as the Gilliland Electrical Manufacturing Company of Indianapolis (AT&T 1958/1979, p. 8). In 1881, Watson, now a wealthy man, left the telephone business to spend a year in Europe and then to pursue another career. In that same year, American Bell bought control of the Western Electric Manufacturing Company of Chicago, reorganized the firm, and transferred the Williams and Gilliland operations to Chicago. In 1882, the Western Electric Company of Illinois became the sole supplier of telephone equipment to American Bell Telephone Company and its associated companies (Lovette 1944–1945, pp. 276–277).

The Western Electric Manufacturing Company had been formed in 1872 by consolidation of Western Union's equipment-manufacturing shop in Ottawa, Illinois, with the firm of Gray and Barton. Elisha Gray was a professional inventor who had devised the first commercially successful telegraphic printer. Enos Barton had been the head operator at Western Union's Rochester headquarters, with the responsibility of examining and testing new inventions in telegraphy. In 1869, Gray and Barton, with financial support from General Anson Stager, superintendent of Western Union, had purchased a shop in Cleveland that made telegraphic instruments. Their firm prospered as manufacturers of fire and burglar alarms and telegraphic and other electrical instruments, several of the most successful of which were of Gray's devising. Gray and Barton then moved their firm to Chicago, where they survived the Great Fire and grew as the city rebuilt. In 1872, Western Union purchased a one-third interest in the company, changed the name, and made the Western Electric Manufacturing Company its instrument supplier. Gray resigned as superintendent in 1874 to return to inventing. (Among his inventions was the "talking telegraph" with which Western Union challenged Bell in 1878.) Barton stayed on and became president of Western Electric in 1886 (Lovette 1944–1945).

For a number of years after Watson's departure, the organization of the technical work at American Bell was very unstable. The Electrical and Patent Department was formed in 1881 by Thomas D. Lockwood, whose attention focused on experiments done for counsel. In 1883, an Experimental Shop was organized under W. W. Jacques to put various patents to practical use. In mid-1884, this Experimental Shop became the Mechanical Department, headed briefly by Ezra T. Gilliland, who had made telephones and switchboards under license from Bell in Indianapolis from 1879 until 1882. In late 1885, following Theodore Vail's departure from American Bell to take charge of the newly formed long-distance company, AT&T, Gilliland resigned, and on December 7, Hammond V. Hayes joined American Bell and took charge of technical development.

In 1891, a second technical department was formed under Joseph P. Davis. The Engineers Department assumed responsibility for the standardization of plant*e* construction and operation. In 1902, the two technical departments were merged to form the Engineering Department, which was run, at least nominally, for three years by a committee. In 1905, Hayes was made chief engineer and placed in full charge.

Finally, in 1907, when Theodore Vail resumed control of AT&T (now the parent company of the Bell System), he fired Hayes, discontinued the Boston Laboratory, and transferred the key people to New York City, to either the Engineering Department of AT&T or the Engineering Department of Western Electric. (Hayes's Enginering Department had remained in Boston even after the assets of American Bell had been transferred to AT&T in New York) (Hill and Shaw 1947; Fagen 1975, pp. 37–44).

Under Hayes's management, as under Watson's, the practical seems to have overwhelmed the theoretical. This occurred despite the fact that Hayes, a Harvard man (class of 1883), had studied electrical engineering at MIT and physics at Harvard, where he was awarded the second Ph.D. in physics granted by that university. Between 1885 and 1907, under Hayes's direction, technical development focused on "fundamental engineering studies, . . . the preparation of standard specifications, data and methods for fundamental planning, bulletins on uniform operating practices, and maintenance instructions" (Hill and Shaw 1947, p. 169).

Some of the earliest fundamental (here "fundamental" means basic to the system, not theoretical) work under Hayes's direction was devoted to finding a more compact and efficient transmission system than the open-wire line. The development of insulated wires with protective coverings (that is, cables) and with ever larger capacities, combined with work on better insulating materials (paper replaced gutta-percha and cotton in the early 1890s), meant that telephone lines could be placed underground in metal conduits. The burial of cables eliminated the towering poles (some were 90 feet high with 30 crossarms) and rooftop racks that had begun to enmesh the larger cities with iron wires and rendered telephone service vulnerable to winds and harsh weather.

Work on transmitters also continued. In 1890, for example, the "solidback" transmitter was developed by Anthony C. White of Hayes's staff. This was the first *commercially* practical design to use multiple contacts, that is, granular carbon, with a fixed rear electrode. Henry Hunnings had received a British patent for a multiple-contact microphonic transmitter using carbon

---

*e*By "plant" is meant both the land the buildings as well as all the machinery, apparatus, and equipment required by an industrial company.

particles in 1878, and in 1886, Edison substituted a better form of carbon. But the problem of electrical or mechanical packing of the carbon granules, which severely shortened the transmitter's life, was left for White to overcome (Hill and Shaw 1947, pp. 158–159; Fagen 1975, pp. 73–83).

But this achievement presented Hayes's staff with new complications. These higher-power microphonic transmitters required more power than human speech alone could supply. This meant that each telephone had to have its own battery cells, which required continual, costly maintenance. The solution of this problem proved to be the common battery system, which substituted one battery at the switchboard or central office for the many on users' premises. As an added benefit, the common battery system also permitted an operator to monitor line use without having to listen in.

But while many practical improvements were being made in telephone equipment and apparatus by trial and error, research was also being done, primarily in Europe, on electrical communication theory (for example, by Lord Kelvin in the 1850s, by James Clerk Maxwell in 1873, by Oliver Heaviside between 1873 and 1901). Because of his academic training, Hayes was one of the few men working in the American telephone industry able to follow the theoretical developments, comprehend their significance, and bring theory to bear upon laboratory work to help understand and possibly to breach physical barriers.

In 1890, on the recommendation of Henry Augustus Rowland, professor of physics at Johns Hopkins, Hayes hired John Stone Stone (Hoddeson 1981, p. 524). Stone was as interested in transmission theory as he was proficient in its practical applications—he was issued some 20 patents during his 10 years in the Boston Laboratory. But perhaps Stone's greatest contribution was the example his work gave, especially to Hayes, of the practical value of theoretical work.

Then, in 1897, Hayes recruited the mathematician George A. Campbell, an MIT graduate who had also studied at Harvard and in Paris, Vienna, and Göttingen (Hoddeson 1981, p. 524). Campbell took up some work of Stone's on the loading of cables (a way of increasing the efficiency of telephone lines by artificially increasing their inductance). By 1899, Campbell had worked out both the theory of loading coils and rules for their design and spacing. Professor Michael I. Pupin of Columbia had also been working on coil-loaded lines and had filed a patent application based on his theoretical solution of the problem that predated Campbell's application by two weeks. (At that point, Campbell's solution was undergoing experimental verification.) The patent application gave Pupin two weeks' priority with respect to theory, although Campbell was the first to verify the theory by experiment (Brittain 1970). The development of inductive loading vastly increased the range and efficiency and reduced the cost of telephone transmission on both open wires and cables, making long-distance telephony feasible.

When John Stone Stone left the Mechanical Department in 1899, he was replaced by E. H. Colpitts and G. W. Pickard. Colpitts, who had studied physics and mathematics at Harvard, assisted Campbell with the development of loading (Hoddeson 1981, p. 526). Pickard, who was educated at Harvard and MIT, followed up on some early work on radiotelephony begun by Stone at Hayes's suggestion in 1892 (Fagen 1975, p. 362; Hoddeson 1981, p. 538). The next (and perhaps the most influential) person to join this cohort of academically trained researchers was Frank Baldwin Jewett, an instructor in physics and electrical engineering at MIT, who was hired in 1904 on Campbell's recommendation (Hoddeson 1981, p. 526). Jewett had a Ph.D. in physics from the University of Chicago, where he had worked under A. A. Michelson and become a friend of Robert Millikan. Thus, a potentially very valuable "Chicago connection" was added to the already profitable "Cambridge connection" that had originated with A. G. Bell himself and had been exploited so successfully by Hammond Hayes. Twenty-one years later, Jewett became the first president of Bell Laboratories.

Hayes's attitude toward the place of theoretical research in industry was complex and ambiguous (Hoddeson 1981, pp. 526–528). On the one hand, in 1892, he wrote to the president of AT&T to explain that he would devote his department's "attention to the practical development of instruments and apparatus," adding that he thought "the theoretical work can be acomplished quite as well, and more economically, by collaboration with the students of the [Massachusetts] Institute of Technology and possibly of Harvard College" (quoted by Reich 1977, p. 19). In 1907, Hayes expressed his reservations about the ability of a scientist to serve as research director:

> The very fact that any great invention at the present must in all probability come from some man of unusual scientific attainments would render a laboratory under the guidance of such men a most expensive and probably unproductive undertaking (quoted by Reich 1977, p. 14).

Yet it was Hayes who had begun to collect men of "unusual scientific attainments" and to put them to work, quite successfully, on tough engineering problems. Hayes was confronting the persistent question of whether creative scientists ought to be promoted, or diverted, into management positions. Does such a promotion mean the loss of a good scientist and the gain of a bad manager? In the Bell System, not only directors of research but presidents of Bell Laboratories have traditionally been selected from among members of technical staff.

One practical result of Hayes's recruitment program was to chip away at the barrier that had existed for some time between the practical "electricians"—the sort who frequented the shop of Charles Williams, Jr.—and the theoreticians. The men collected by Hayes formed a solid core of creativity for the Bell System.

In 1907, almost as if he were taking Hayes at his word, Vail replaced him with a man who had no academic credentials. The new chief engineer of AT&T was John J. Carty, originally from Cambridge, Massachusetts. Carty had begun his telephone career in 1879 as one of the original boy operators, when eye trouble had precluded his enrolling at Harvard. Carty then joined the New England Telephone Company, where he became involved in technical work, introducing into commercial use the full metallic circuit multiple switchboard, which used a common battery. Between 1883 and 1896, Carty was issued 24 patents. In 1887, Carty briefly joined Western Electric's cable department, where he supervised the laying of cable in cities in the eastern United States and worked successfully on the problem of neutralizing inductive disturbances, which were inhibiting the expansion of long-distance service. Two years later, he became "Electrician," or chief engineer of the Metropolitan Telephone and Telegraph Company (a precursor of the New York Telephone Company). In New York, Carty reorganized the technical staff and began recruiting graduates of scientific and engineering schools. This had not been done before by any of the Bell System operating companies. Carty also modernized the switchboard and cable plant—metallic or 2-wire circuits were introduced and overhead open wires were replaced by underground cables—and upgraded service with the introduction of new traffic, equipment, and construction methods (Bell Laboratories 1930, p. 16).

Vail saw in Carty a man of both vision and action. For example, in 1906, in a paper entitled "Telephone Engineering," read before the American Institute of Electrical Engineers, Carty had argued that the telephone engineer must place himself within the organization so as to be able to coordinate and "fairly consider all of the projects and ideas pertaining to the design, operation, construction and maintenance of the plant. . . . But," he continued, "what is still more important, the successful engineering of a telephone plant depends upon proper business management. . . . Without an intelligent, progressive and broad-gauged business management, there cannot be good telephone engineering" (Jewett 1937, pp. 168–169).

One of Carty's enduring visions was of a telephone system that spanned the world, to "join all the people of the earth into one brotherhood" (Jewett 1937, p. 175). To this end, as chief engineer of AT&T from 1907 until 1919, Carty oversaw the development of the first transcontinental telephone line (which opened between New York and San Francisco on 25 January 1915) as well as the first trans-Atlantic radiotelephone transmission (on 21 October of that same year between the U. S. Naval Station at Arlington, Virginia, and the Eiffel Tower in Paris).

As World War I threatened to involve the United States, both the army and the navy called on Carty to supply them with the most effective communications systems. Carty mobilized the Bell System, both people (12 battalions) and equipment, to create what Colonel (later General) Salzman, acting chief signal officer, was to describe as "a wonderful system of communication of an efficiency and capacity never contemplated in the history of warfare" (Jewett 1937, p. 173). In 1921, for his war service, Carty was created a brigadier general in the Officers' Reserve Corps (and was thereafter always referred to as "General Carty").

Perhaps Carty's most far-reaching contribution to the Bell System—and to telecommunications—was his vision of the role science might play in an industry that was dependent upon an increasingly complex technology. Carty became both an initiator of industrial research and one of its most forceful advocates (Jewett 1937, p. 163). He was a well-known speaker and writer on "electrical subjects." He encouraged universities to undertake the kind of scientific research whose results might find applications in industry and, in turn, encouraged industry to support university research in pure science. I have already mentioned the fact that, when Carty was appointed chief engineer of AT&T in 1907, he reorganized the research and development effort in the Bell System, consolidating in New York City the laboratory staffs from Boston, Chicago, New York, and elsewhere. In 1911, a separate Research Branch was formally organized within the Engineering Department of Western Electric, with what Carty described as "the best talent available and . . . the best facilities possible for the highest grade laboratory work" (quoted by Fagen 1975, p. 44). E. H. Colpitts was put in charge, reporting to Assistant Chief Engineer F. B. Jewett.

The new Research Branch was immediately put to the test. Carty's objective of transcontinental service had been given a time frame. On a visit Vail, Carty, and Bancroft Gherardi (plant engineer of AT&T) made to San Francisco in 1909, they promised to have such service in place for the opening of the Panama-Pacific Exposition scheduled for San Francisco in 1914. (The opening was postponed until 1915.) Honoring this commitment required the solution of a very serious technical problem, which in its turn had an immense impact on the direction of research in the Bell System as well as upon telecommunications as a whole.

Even with loading, Denver seemed to be the farthest west (2100 miles) that long-distance lines could reach from New York. Ohmic attenuation, the dissipation of energy, set a natural limit that could not be overcome by further engineering improvements of the already-known technology, mechanical amplification. Jewett suspected that the answer might be found in the "electron streams" with which his friend Robert Millikan had been "playing" for years. In response to Jewett's request for someone trained in the "new physics" who might probe the problem further, Millikan recommended H. D. Arnold. Arnold joined the Engineering Department at Western Electric in 1911, having completed his doctoral studies at the University of Chicago.

Late in 1912, with the assistance of John Stone Stone (who had left the Bell System in 1899), Lee De Forest arranged to give an audience of Bell System technical people a demonstration of how a device he had invented in 1906–1907 could be used as an amplifier. De Forest's device, a 3-element, low-vacuum tube, performed erratically, but well enough to convince Arnold of its possibilities—if ionizable gases could be evacuated from the tube.

Arnold dropped the line of investigation he had been following with some successes (mercury-vapor discharge tubes adapted from earlier work by Peter Cooper Hewitt). Within a year, he succeeded in evacuating the tube of the audion quite well enough to produce commercially successful vacuum-tube amplifiers. These were installed on the first transcontinental line, which was tested on 29 July 1914 (AT&T 1958/1979, p. 20). Arnold did not have the field to himself. Irving Langmuir of General Electric's research laboratories was also working with high-vacuum tubes at this time. A long patent contest was finally resolved in favor of Arnold's view that no new invention was actually involved. Arnold's initial contribution chiefly consisted of (1) recognizing the necessity of a high vacuum, and (2) developing practical methods to achieve one (AT&T 1958/1979, p. 19).

An important result of Arnold's development of the electronic amplifier was the beginning, in 1912, of a program of research and development at both Western Electric and AT&T on the physics and chemistry of electron emission—to understand thoroughly the phenomena involved and to provide improved designs of tubes and circuits (Fagen 1975, pp. 967–971).

The vacuum tube, improved by Arnold and others, turned out to be a versatile device. It evolved into a virtually distortion-free repeater for voice frequencies, an oscillator, detector, rectifier, modulator, and demodulator. In fact, until the advent of solid-state devices, the vacuum tube was essential for long-distance communication by wire and cable, by radio and television. In the ENIAC, it also played a role in the birth of computers.

In 1919, the Engineering Department of AT&T was divided into two new departments: Development and Research under J. J. Carty, now a vice-president, and Operations and Engineering under N. C. Kingsbury. By 1924, technical work in the Bell System had expanded in both scope and intensity, and yet another reorganization seemed necessary. On 27 December 1924, the Engineering Department of Western Electric became a separate corporation entity, Bell Telephone Laboratories, Incorporated, which began operations on 1 January 1925.

Bell Laboratories was chartered to perform scientific and engineering research and development for AT&T and to incorporate this research into designs for Western Electric. General Carty became chairman of the board of Bell Laboratories and F. B. Jewett, its first president. E. B. Craft, formerly chief engineer of Western Electric, became executive vice-president and H. D. Arnold, director of research. Finally, in 1934, the Development and Research Department of AT&T was consolidated into Bell Laboratories.

With the formation of Bell Laboratories, the formative era of research in the Bell System ended, and the modern industrial laboratory achieved its familiar form. Its growth was clearly in response to the technological needs of a rapidly expanding telecommunications system and the result of the faith of certain individuals that technological development would benefit from an in-house understanding of modern scientific methods and theories. Over and over again, this faith was justified, as for example, when Campbell's knowledge of mathematics led to loading coils and when Arnold's training in theoretical physics led to the vacuum tube.

THE ESTABLISHMENT OF THE BELL LABORATORIES ARCHIVES

In July of 1980, with the appointment of an archivist, Bell Laboratories officially established an archives. In July of 1981, a small archival storage and research facility was opened in a new building in Short Hills, New Jersey.

The need for such a facility has been widely recognized for some time. Old Bell System records have proven their value for both legal and historical research—much of it prompted by the volumes of the Bell System history already published. But such research has not been as efficient as it might be, because of the lack of proper archival research facilities and finding aids. Because research, development, and support activities have been expanding at Bell Laboratories, storage space is at a premium. What is not in use is usually recycled or disposed of, while what is preserved for future use is "warehoused." Many (but not all) noncurrent records, such as laboratory notebooks, fall into the latter category. Under unsuitable storage conditions and unsupervised use, a slow but inevitable destruction of records usually takes place as a result of fluctuating temperature and humidity and the accumulation of dirt as well as folding and unfolding, jostling, and other mishandling.

Records management schedules, moreover, provide for the disposal of certain classes of records at stated times—after 30 years, for example, for laboratory notebooks. Fortunately, the foresight of certain people has delayed the execution of this schedule, pending decisions about the historical value of the materials. But people with an appreciation for the historical value of noncurrent records are few and far between, and preservation cannot be left to chance. Archival and historical values should be built into records management schedules.

Establishing an archives in an industrial laboratory is not without its problems. For one thing, a research organization naturally tends to look forward rather than backward over its achievements. The technical library at Bell Laboratories, for example, tries to keep the most up-to-date information and is loathe to give valuable shelf space to out-of-date journals and books. Further, there is the issue of financial responsibility and accountability—not only with regard to our management in Bell Laboratories and to our owners, AT&T and Western Electric, but also to ratepayers and various regulatory agencies. Public utilities commissions are understandably reluctant to spend ratepayers' money on what many people think are unnecessary enterprises. The latter may include anything not specifically project oriented, for example, research in both basic science and history. Then there is the fact that not all the records an archivist or historian is interested in find their way into the central files. Those records that might be needed to support a patent application are very well provided for, laboratory notebooks and case files, for example. But records of an administrative nature, on the department, laboratory, or personal level, are not so well protected, especially during personnel changes. Artifacts fare especially badly. At Bell Laboratories, equipment is often dismantled and recycled—as are the modular laboratories themselves. (So you will search in vain for the actual room in which the transistor was first demonstrated.) Sometimes early crystals, transistors, vacuum tubes, or sections of cables survive only as personal mementos.

Modern technology, with all its benefits, can also cause new difficulties. While the telephone reduced the number (and quality) of letters, memos, telegrams, etc., that an archives might otherwise have inherited, copying machines threaten to bury us all under stacks of copies. Proliferating computer terminals, with their machine-readable records and memories, pose another sort of threat (which I cannot go into here). At Bell Laboratories, the number of laboratory notebooks being issued has already dropped off, even though the technical staff is growing in numbers.

None of these problems is unique to Bell Laboratories, of course. In fact, there are probably far fewer difficulties to overcome at Bell than at other industrial institutions or corporations. Some companies, after all, are reported to have surrendered to their fear of litigation and mandated the destruction of all company records within two or three years of their creation.

But whatever the problems may be in organizing an archives at Bell Laboratories, they are more than compensated for by assets. People at every level have been cooperative and even enthusiastic about the project. Everyone is aware of Bell Laboratories's superb record of achievement, and some have seen the need to document it. In particular, there is a small but dedicated staff of people who have been working with historical documents for some time now, but who have been at the Bell Laboratories even longer and so know the company and its personnel very well indeed.

But the greatest asset is the core of documents the archives has inherited. Some have already been transferred to the new storage facility; some are still in a warehouse.

- Over 100,000 laboratory notebooks and thousands of case files, which document research and development activities
- The so-called Boston Files, which were brought to New York when the headquarters of the Bell System moved from Boston
- The executive files, which comprise the papers of Carty, Jewett, Arnold, and Colpitts, among others
- Papers written and collected by some Bell System historians, including Lloyd Espenschied, Roger B. Hill, and Morton Fagen, the editor of the first two volumes of the Bell System history
- A collection, partially transcribed, of oral histories
- Early mono and stereo record masters, some of which were recently transcribed to produce two albums of selections by Leopold Stokowski and the Philadelphia Orchestra
- Some recordings of very early radio broadcasts
- Some documents, photographs, and negatives from the Bell System Historical Museum, which are the remains of an earlier attempt to preserve an historical record of Bell System accomplishments.[f]

There is great diversity and a vast quantity of material, most of which is accessible only with some difficulty. The establishment of the archives will ensure the preservation and expansion of these resources, make them accessible, and assist Bell System researchers and other qualified scholars in their use.

[f] J. J. Carty authorized the establishment of the Bell System Historical Museum in late 1912. It began operation in 1913, with the appointment of Wilton L. Richards (who had begun his career in the Bell System as assistant to Emile Berliner) as curator. In 1922, the museum, which was located in Western Electric's headquarters at 463 West Street, was supplemented by what became the American Telephone Historical Library. The latter, the William Chauncy Langdon as historical librarian, was located in AT&T's headquarters at 195 Broadway (Farnell 1936; Langdon 1924–1925).

CONCLUSION

As the nineteenth century waned and the twentieth century began, the Bell System underwent rapid growth, heavily punctuated by crises and reorganizations. My narrative has suggested that the changes it underwent and the shape it took were the result of both action and reaction. On the one hand, there were the vision and initiative of individuals: to spend profits on expansion, to commit the company to nation-wide and even world-wide service, to spend money on research, to hire academically trained scientists as well as engineers. On the other hand, there were reactions to outside forces: to legislative limits on refinancing, to the physical limits of old technologies, to competition, to patent challenges. The outcome of all this was a corporate structure heavily dependent upon and committed to a complex technology supported by in-house research and development.

I have also traced the growing importance to the Bell System of research based upon an understanding of the latest scientific developments. As a consequence, an organization was created, drawing its talent from the finest academic institutions—as well as from among the naturally talented— capable of performing mission-oriented research and development. The present embodiment of that organization, the Bell Laboratories, is designed by and for technological challenges. As a continuously interacting scientific and technical operation, it unites investigations at the forefront of those areas of pure science significant to telecommunications with the technological development, the design, and the engineering necessary to manufacture new products or devices.

Finally, I have briefly described some of the resources currently being made available to scholars interested in knowing more about the origins, development, and achievements of Bell Laboratories.

REFERENCES

AT&T. 1958/1979. Events in Telecommunications History. New York. This chronology was first published in 1958 and is updated regularly. Page references are to the 1979 version.

BELL LABORATORIES. 1930. John J. Carty—a Biographical Note. Bell Laboratories Record 9: 14–19.

BELL LABORATORIES. 1933. Harold de Forest Arnold. Bell Laboratories Record 11: 350–360.

BRITTAIN, J. E. 1970. The Introduction of the Loading Coil: George A. Campbell and Michael I. Pupin. Technology and Culture 11, 36–57.

BUCKLEY, O. 1952. Frank Baldwin Jewett, 1879–1949. National Academy Biographical Memoirs 27: 239–264.

FAGEN, M. D., Ed. 1975. A History of Engineering and Science in the Bell System. Vol. 1: The Early Years (1875–1925). Bell Telephone Laboratories, Incorporated. Murray Hill, New Jersey.

FARNELL, W. C. F. 1936. The Bell System Historical Museum. Bell Telephone Quarterly 15: 169–187 and 261–273.

HILL, R. B. & T. SHAW. 1947. Hammond Hayes: 1860–1947. Bell Telephone Magazine 26: 150–173.

HODDESON, L. H. 1981. The Emergence of Basic Research in the Bell Telephone System, 1875–1915. Technology and Culture **22:** 512–544.

HOUSHELL, D. A. 1975. Elisha Gray and the Telephone: On the Disadvantages of Being an Expert. Technology and Culture **16:** 133–161.

JEWETT, F. B. 1937. John J. Carty: Telephone Engineer. Bell Telephone Quarterly **16:** 160–177.

LANGDON, W. C. 1923. The Early Corporate Development of the Telephone. Bell Telephone Quarterly **2:** 133–152.

LANGDON, W. C. 1924–1925. The American Telephone Historical Collection. Bell Telephone Quarterly **3:** 41–49; **4:** 143–156.

LOCKWOOD, T. D. 1926. Stories by an Early Pioneer. Telephone Topics **19:** 495–500. Special issue of a New England Telephone Company employee magazine celebrating the fiftieth anniversary of the telephone.

LOVETTE, F. H. 1944–1945. Western Electric's First 75 Years: A Chronology. Bell Telephone Magazine **23:** 271–287.

MABON, P. C. 1975. Mission Communications: The Story of Bell Laboratories. Bell Telephone Laboratories, Inc. Murray Hill, N. J.

NOBLE, D. F. 1977. America by Design: Science, Technology, and the Rise of Corporate Capitalism. Alfred A. Knopf. New York.

REICH, L. S. 1977. Radio Electronics and the Development of Industrial Research in the Bell System. Unpublished Ph. D. dissertation, Johns Hopkins University.

SOLOMON, R. J. 1978. What Happened after Bell Spilled the Acid? Telecommunications Policy **2:** 146–157.

WATSON, T. A. 1913/1979. The Birth and Babyhood of the Telephone. An address delivered before the 3rd Annual Convention of the Telephone Pioneers of America at Chicago, October 17, 1913. Reprint, 1979, AT&T, New York. Page references are to the AT&T brochure.

Contextualism and the Character of Developmental Psychology in the 1970s[a]

RICHARD M. LERNER,[b] DAVID F. HULTSCH, AND
ROGER A. DIXON[c]

Department of Individual and Family Studies
The Pennsylvania State University
University Park, Pennsylvania 16802

The purposes of this paper are, first, to describe, explain, and illustrate a major alteration in the philosophy of science underlying theory and research in developmental psychology in the 1970s. We will argue that contextualism both transcended and integrated the organismic and the mechanistic models of development that dominated developmental psychology in preceding decades. Second, we will provide examples of how contextualism offers a useful model for lower order theoretical, methodological, and empirical concerns.

Our discussion will proceed by reviewing the history of three central themes in developmental psychology in the 1970s. These themes are: first, the relation between philosophical paradigms or world views and science; second, the life-span, multidisciplinary view of human development; and third, the nature-nurture issue, or more broadly, the issue of the character of organism-environment relations. In our view, the synthesis of these three themes led to the emergence of contextualism as a major organizational philosophy within developmental psychology during the 1970s. Indeed, as we will illustrate, it is a contextual point of view which itself allows one to understand this history. The influence that each theme had on developmental psychology is best understood in the context of the others. In other words, each theme was both a product and a producer of the other two.

Depicting the philosophical paradigms that characterized an area within psychology over the course of an entire decade is always difficult. As noted by Royce (1976), psychology is multiparadigmatic, and thus, any choice of a single or limited set of paradigms is likely to reflect the unique viewpoint of the historian. Nevertheless, such accounts are possible and often fruitful (e.g., see Bronfenbrenner, 1963; Looft, 1972; Riegel, 1973). They draw our attention to the nonempirical assumptions that have determined the theoretical and methodological concerns of science. Such accounts also help us understand the

[a]This paper was presented at the meeting of the Section of History, Philosophy and Ethical Issues of Science and Technology of The New York Academy of Sciences held on April 23, 1980.

[b]Richard M. Lerner's work on this manuscript was supported in part by a grant from the John D. and Catherine T. MacArthur Foundation.

[c]Now at the Max-Planck Institute for Human Development and Education, Berlin, West Germany.

0077-8923/83/0412-0101 $01.75/2 © 1983, NYAS

nature of the current philosophical-empirical linkages guiding science and may serve as guides for prospections about the future. These contributions may be especially useful when a given account capitalizes on the perspective gained by the passage of time and thus the accumulation of multiple sources of evidence.

In the present paper, however, we make our argument with the corpus of the preceding decade not completely cold. We are aware of the benefit of a greater temporal distance in providing the opportunity for the emergence of evidence buttressing our views; in turn, we know of the limits of writing history about the historical moment within which one is embedded. But, to a contextualist, the historical moment is the basic datum of science. In fact, the contextual paradigm, as outlined by Pepper (1942), sees history as an attempt to present again, or re-present, an event (Sarbin, 1977). Thus, history is not an event necessarily in the past; rather, in contextualism the event is alive and in the present (Sarbin, 1977). As a consequence, history is created anew each time it is written, and the perspective one seemingly gains with a greater temporal distance from the target event is, from a contextual perspective, not an aid to better presenting the past: rather, it is a moderator for our constructions of the present.

In our view, this prefatory contextual analysis, or defense if you wish, of writing the history of contextualism in the development psychology of the 1970s so early in the 1980s, is not just a self-reinforcing conceptual tool. As we will seek to illustrate, this notion that the historical event provides the basic datum of science leads to a framework for understanding some novel and exciting research that arose in the 1970s within the areas of cognitive development, on the one hand, and personality and social development, on the other. For example, in a later portion of the paper we will elaborate on the notion that, like history, the contents of memory are changed each time they are employed (e.g., see Hultsch & Pentz, 1980; Meacham, 1977). We will argue that such contextual ideas offer provocative ways of extending research into the 1980s, not only for the further elaboration of description and explanation of developmental processes, but also for the derivation of new strategies for the optimization of human life. To present these assets, however, we must first consider the nature of, and interrelations among, the three themes we have suggested as characterizing the 1970s.

CONCEPTUAL THEMES IN DEVELOPMENTAL PSYCHOLOGY IN THE 1970s

The disciplines involved in the study of human development themselves undergo developmental change (Hartup, 1978; Riegel, 1972). In developmental psychology, the decades preceding 1970–1979 were characterized by a movement away from the descriptive and normative study of human development (Sears, 1975) to a primary focus on process and explanation (Bronfenbrenner, 1963; Looft, 1972). Characterizing this emphasis Mussen (1970, p. vii) noted, "The major contemporary empirical and theoretical emphasis in the field of developmental psychology, however, seems to be on explanations of the psychological changes that occur, the mechanisms and processes accounting for growth and development." The 1970s saw an increasingly more abstract concern with the study of development, and thus the first of three interrelated themes that emerged involved a renewed appreciation of the metatheoretical, or paradigmatic, bases of developmental theories (Overton & Reese, 1973; Reese & Overton, 1970).

The Role of Paradigms of Human Development

In the 1970s discussions occurred about how the "family of theories" (Reese & Overton, 1970) derived from the organismic and the mechanistic world views, respectively, were associated with different ideas about the nature and nurture bases of development (Overton, 1973); with different stances in regard to an array of key conceptual issues of development, e.g., the quality, openness, and continuity of change (Looft, 1973); with contrasting methodologies for studying development (Overton & Reese, 1973); and, ultimately, with alternative truth criteria for establishing the "facts" of development (Overton & Reese, 1973; Reese & Overton, 1970).

This focus on the philosophical bases of developmental theory, method, and data analysis led to considering the use of still other paradigms in the study of development (e.g., Riegel, 1973). In part, this concern occurred as a consequence of interest in integrating assumptions associated with prototypic organismic- and mechanistic-derived theories (Looft, 1973). For instance, Riegel (1973, 1975, 1976a, 1976b) attempted to formulate a paradigm of development including both the active organism focus in organicism and the active environment focus in mechanism. In addition, however, interest in continual, reciprocal *relations* between active organism and active context (and not in either "element" per se), and concern with these relations as they existed on all phenomenal levels of analysis, was a basis of proposing a dialectical (Riegel, 1975, 1976a), transactional (Sameroff, 1975), or relational (Looft, 1973) model of human development.

Contrary to assumptions associated with the other paradigms of human development, Riegel's (1975, 1976a, 1976b) view emphasized the continual conflict among inner-biological, individual-psychological, outer-physical, and sociocultural levels of analysis; he assumed that constant changes among the multiple, reciprocally related levels of analysis were involved in development (Overton, 1978). Thus, at least at this level of analysis his dialectical model can be seen as compatible with the paradigm that Pepper (1942) labeled as *contextualism* (Hultsch & Hickey, 1978; Lerner, Skinner & Sorell, 1980). From this perspective, developmental changes occur as a consequence of reciprocal (bidirectional) relations between the active organism and the active context. Just as the context changes the individual, the individual changes the context. As such, by acting to change a source of their own development, by being both a product and a producer of their context, individuals affect their own development (Riegel, 1976a; see also Schneirla, 1957). Moreover, since the context is seen as a multilevel one—having interrelated biological, sociocultural, physical environmental, and historical components—Riegel's ideas meshed with the second theme that can be identified as emerging in the study of developmental psychology in the 1970s. That is, by stressing that there is change across the life-span, and that it involves variables from several levels of analysis, Riegel's ideas were compatible with a multidisciplinary, life-span view of human development.

The Multidisciplinary, Life-Span Study of Human Development

Developmental psychology is not the only scientific discipline concerned with the study of human development. Family and life-course sociologists, developmental and evolutionary biologists, comparative psychologists, physicians, and economists are also concerned with human development (e.g., see Riley, 1979). Some reviewers (e.g., Hartup, 1978) have noted that despite the relative disciplinary isolation that characterized research prior to the 1970s, the years following this time were marked by calls for interdisciplinary integration (e.g., Bronfenbrenner, 1977; Burgess & Huston, 1979; Hill & Mattessich, 1979; Lerner & Spanier, 1978; Petrinovich, 1979; Riley, 1979; Wilson, 1975). The bases for these calls were primarily conceptual. Although it is possible to present these bases from multiple disciplinary foci (e.g., from a life course sociological perspective; Riley, 1979), the emphasis on changes in the psychological analysis of human development is a useful representative case.

Attempts to use a unidimensional biological model of growth, based on an idealistic, genetic-maturational (organismic) paradigm, to account for data sets pertinent to the adult and aged years were not completely successful (Baltes, Reese, & Lipsitt, 1980; Baltes & Schaie, 1973). Viewed from the perspective of this organismic conception, the adult and aged years were necessarily seen as periods of decline. However, all data sets pertinent to age changes, e.g., in regard to intellectual performance, during these periods were not consistent with such a unidirectional format for change. Increasingly greater interindividual differences in intraindividual change were evident (Baltes, 1979a; Baltes & Schaie, 1974, 1976; Schaie, Labouvie, & Buech, 1973). Variables associated with membership in particular birth cohorts and/or with normative and nonnormative events occurring at particular times of measurement appeared to account for more of the variance in behavior change processes with respect to adult intellectual development than did age-associated influences (Baltes et al., 1980). Data sets pertinent to the child (Baltes, Baltes, & Reinert, 1970) and the adolescent (Nesselroade & Baltes, 1974) that considered these cohort and time effects also confirmed their saliency in developmental change. Conceptualizations useful for understanding the role of these non-age-related variables in development were induced (e.g., Baltes, Cornelius, & Nesselroade, 1977).

As a consequence of this empirical and conceptual activity, the point of view labeled as "life-span developmental psychology" or as the "life-span view of human development" (Baltes, 1979a, 1979b; Baltes et al., 1980) became crystallized. The emerging nature of this orientation has become clear over the course of several conferences (Baltes & Schaie, 19773; Datan & Ginsberg, 1975; Datan & Reese, 1977; Goulet & Baltes, 1970; Nesselroade & Reese, 1973), the initiation of publication of an annual volume devoted to life-span development (Baltes, 1978; Baltes & Brim, 1979), and the publication of numerous empirical and theoretical papers (Baltes *et al.,* 1980). From this perspective, the potential for developmental change is seen to be present across all of life; the human life course is held to be potentially multidirectional and necessarily multidimensional (Baltes, 1979b; Baltes & Nesselroade, 1973; Baltes *et al.,* 1980). In addition, the sources of the potentially continual changes across life are seen to involve both the inner-biological and outer-ecological levels of the context within which the organism is embedded. Indeed, although an orientation *to* the study of development and not a specific theory *of* development (Baltes, 1979b), it is clear that life-span development-alists are disposed to a reciprocal model of organism-context relations. As Baltes (1979b, p. 2) has indicated, there are two rationales for this contextual emphasis:

> One is, of course, evident also in current child development work. As develop-ment unfolds, it becomes more and more apparent that individuals act on the environment and produce novel behavior outcomes, thereby making the active and selective nature of human beings of paramount importance. Furthermore, the recognition of the interplay between age-graded, history-graded, and nonnormative life events suggests a contextualistic and dialectical conception of development. This dialectic is further accentuated by the fact that individual development is the reflection of multiple forces that are not always in synergism, or convergence, nor do they always permit the delineation of a specific set of endstates.

In sum, the development of life-span developmental psychology in the 1970s led to a multidisciplinary view of human development, one suggesting that individual changes across life are both a product and a producer of the multiple levels of context within which the person is embedded.

It is important to note three points about the present status of this view. First, in order to study the complex interrelations among organism and context life-span developmentalists promote the use of particular research designs and methodologies (e.g., sequential designs, multivariate statistics, cohort analysis). Second, they seek both methodological and substantive collaboration with scholars from disciplines whose units of analysis have traditionally been other than individual-psychological, or personological, ones. For example, the work of life course sociologists has been important in advancing life-span development psychology (e.g., Elder, 1974, 1979; Riley, 1979). Third, however, these methodological and multidisciplinary activities are undertaken primarily for conceptual reasons. If contextual influences were not seen as crucial for understanding individual development, then neither methods for their assessment in relation to the individual, nor information about the character of these levels of anlaysis, would be necessary.

Accordingly, the life-span view promotes a model of development we have seen described as a contextual one (Pepper, 1942). In so doing, it sees individuals as both products and producers of the context which provides a basis of their development. As such, individuals may be seen as producers of their development. This idea of reciprocal relations between individuals and their environment relates to the third theme that emerged in the 1970s: the reconceptualization of the nature-nurture issue.

The Nature-Nurture Issue

For decades there has been a general recognition that heredity and environment, nature and nurture, are both needed for behavior, but confusion and controversy persist in regard to how processes from each provide the source of behavior. In 1958, Anastasi suggested that many psychologists continued to ask the wrong question regarding nature and nurture. They asked "how much of each?" source was required for a given behavior. Anastasi (1958) persuasively argued that 100% of each was necessary. There would be, after all, no place to see the effects of heredity without environment, and there would not be anyone in the environment without heredity. Thus, she rejected the "how much of each?" formulation of the issue, and advanced the idea that the fundamental question is how the two sources, both completely present, interact to provide the basis of behavior. Although she was joined in this idea by others (e.g., Hebb, 1949, 1958; Lehrman, 1953, 1970; Schneirla, 1957), the nature-nurture issue regained marked prominence in the late 1960s precisely in the terms of the "how much of each?" question.

The ideas of Jensen (1969) regarding the heritability of intelligence implied for many that a useful index had been found to measure the magnitude of hereditary contribution to a trait. However, heritability is a statistic *describing* a property of a specific distribution at a specific point in time (Hirsch, 1970; Kamin, 1974; Layzer, 1974; Lewontin, 1976). It thus does not reflect a measure of an intraindividual characteristic, but only of interindividual differences. At best, heritability is an index of the degree to which interindividual differences in a trait distribution can be accounted for by interindividual differences in genotypes (Hirsch, 1970; Lerner, 1976). However, many people, although not necessarily Jensen (e.g., see Jensen, 1973), made a mistaken interpretation of its meaning and argued that if heritability were high then: (1) the trait being measured was genetically determined; and (2) there was little hope of environmental intervention modifying the trait.

Basically, there were two problems involved in these inferences. At a relatively empirical level was the view that on the basis of a group statistic, that described the extent to which differences *between* people could be accounted for by reference to genotype differences between them, one could infer how much of a trait *within* a given person was determined solely by heredity (see Lehrman, 1970; Lerner, 1976). Although portions of Jensen's (1969, 1973) own writings spoke against this belief, and others also pointed out the problems with the reasoning (e.g., Feldman & Lewontin, 1975; Hebb, 1970; Kamin, 1974; Layzer, 1974; Lewontin, 1976), several people maintained these views (e.g., Eysenck, 1971; Herrnstein, 1971).

However, the second, and in our view more important, problem was the metatheoretical dispute among participants in the controversy (cf. Baltes, Reese & Nesselroade, 1977). Those who favored the use of heritability coefficients were essentially arguing from a mechanistic, genetic reductionist viewpoint (see Overton, 1973). This perspective led them to view hereditary and environmental variance as linearly additive and separable components of behavior functions. In turn, those who rejected this analysis of variance analogue to the conceptualization of nature and nurture (e.g., Hebb, 1970), argued from a metatheoretical stance that stresses strong, or dynamic, interactions between heredity and environment (e.g., Lerner, 1976, 1978; Overton, 1973), and thus a viewpoint consonant with the contextual and life-span themes already discussed.

Accordingly, because the core dispute about the heritability of intelligence was paradigmatic, little light could be cast by focusing on the extant empirical issues. As such, a refired nature-nurture controversy characterized the early 1970s. Moreover, given the fact that Jensen (1969) made inferences about the relevance of heritability scores in accounting for racial differences in IQ scores, his "genetic differences" hypothesis (Jensen, 1969) was soon termed by others a "genetic inferiority" hypothesis (cf. Scarr-Salapatek, 1971). This gave the issue important social implications. Additionally, because some argued that if the heritability of IQ was high, then social policies or education programs designed to enhance the intellectual performance of individuals would fail (see Jensen, 1973 for a critique of this argument), this societal impact of the debate was enhanced.

Furthermore, the renewal of controversy about the nature and nurture of intelligence seemingly opened up a Pandora's Box of concern regarding the relative contributions of nature and nurture to sexism, militarism, social Darwinism, racism (Tobach *et al.*, 1974), educability (Jensen, 1973), and sex differences in personality (e.g., Carlson, 1972), to name just some of the areas of concern. Indeed, as evidenced by recent contributions to the literatures of several disciplines (e.g., Feldman & Lewontin, 1976; Loehlin, Lindzey & Spuhler, 1975; Lewontin, 1976; Wilson, 1975) not only has the debate regarding the contributions of nature and nurture to human functioning not been resolved to date, but instead, it has evolved as a concern having multidisciplinary relevance. Perhaps the best example of the multidisciplinary dimensions of this debate arose in 1975 with the publication of E. O. Wilson's *Sociobiology: The New Synthesis*.

Sociobiology, as promoted by Wilson and others (e.g., Trivers), attempts to integrate through biological reductionism not only the biological sciences, but the social sciences and the humanities as well. As noted by the philosopher Caplan (1978, p. 2) this approach is "the latest and most strident of a series of efforts in the biological sciences to direct scientific and humanistic attention toward the question of what is, fundamentally, the nature of human nature." Consistent with the metatheoretical assumptions we have seen associated with those promoting the use of heritability coefficients, many sociobiologists construe nature as a preformed, immutable contribution to behavior. That is, whatever the proportion of variance in human social behavior with a genetic basis, it is that proportion that is genetically constrained and generally unavailable to contextual influence.

The criticisms of sociobiology have come from the several disciplinary quarters that sociobiologists seek to digest (e.g., see Caplan, 1978). Within the biological and social sciences the criticisms have generally been associated with conceptualizations stressing that sociobiologists do not appreciate the plasticity of genes, organisms, or contexts, and that just as genes influence their contexts the reverse is also the case. Again then, metatheoretical division, akin to the one dividing those who do or do not see the use of heritability calculations, characterizes this instance of the nature-nurture debate of the 1970s. Indeed, at least insofar as developmental psychology is concerned, the stress on the reciprocal character of organism-environment relations promoted by a contextual, life-span perspective, has led to general disfavor with sociobiological thinking, and in fact with any perspective that does not see human development as a quite plastic phenomenon arising out of a dynamic interaction between nature and nurture (Baltes & Baltes, 1979c; Lerner, 1978; Overton, 1973; Sameroff, 1975).

Conclusions

To summarize the preceding discussion, we may note that (1) after decades of increasing interest in developmental theory, further conceptual elaboration resulted in a more abstract plane of discussion, one that acknowledged the role of paradigms and therefore the philosophical bases of theories; (2) developments in science have been associated with the elaboration of a multidisciplinary, life-span view of human development; (3) because of these developments the 1970s involved: (a) a concern with a contextual model of development; (b) a concern with the idea that individuals, in action with their changing context, can provide a basis of their own development; and (c) the view that nature and nurture, organism and environment, are reciprocally acting bases of human development, and that, as such, the potential plasticity of human development is assured.

Thus, each of the three themes we see as characterizing the history of developmental psychology in the 1970s was both a product and a producer of the other two; and, similarly, the contextualism that gives meaning to this confluence was also the outcome of it. Indeed, throughout the 1970s there were numerous calls for promoting contextualism as the paradigm of psychology in general, and developmental psychology in particular. As noted earlier, these calls suggested that contextualism offers an empirically useful means for advancing the scientist's ability to describe, explain, and optimize human functioning in ecologically meaningful settings. These appeals may be as much a component of our present and future as they are of our past. As such, we need to depict and evaluate their content in order to understand the role the contextual paradigm played in the 1970s and the contribution it is likely to be associated with in the 1980s. To do this we will first, briefly, review the assumptions of a contextual paradigm. Second, we consider the nature of the appeals for the deployment of this paradigm. Third, we explore the empirical potential of contextualism by briefly considering the application of contextually-derived theorizing to two research literatures. Finally, we conclude with our prospections about the future use of contextualism in theory, research, and the optimization of human life.

ASSUMPTIONS OF A CONTEXTUAL PARADIGM

Over the course of its history, knowledge about psychological development has been advanced most notably by research derived from either an organismic or mechanistic paradigm (Baltes, 1979b; Overton & Reese, 1973; Reese & Overton, 1970). Although contributions derived from the two paradigms have remained relatively unintegrated (Kuhn, 1978), both paradigms nevertheless share some common assumptions (cf. Kaufmann, 1968). These assumptions pertain to the views that: (1) the universe is uniform and permanent; and (2) that the laws to be discovered about the development of organisms are all absolute ones, whether they involve variables lying inside the organism (for example, arising as a product of its nature) or outside the organism (for example, in its nurture). Although sampling and technological limitations lead science to be able to generate only probabilistic laws, the probabilistic statements are regarded as unbiased estimates of absolute ones (Hempel, 1966).

However, a contextual paradigm assumes: (1) *constant change* of all levels of analysis, and (2) *embeddedness* of each level with all others, that changes in one promote changes in all. The assumption of constant change denotes that there is no complete uniformity or constancy. Rather than change being a to-be-explained phenomenon, a perturbation in a stable system, change is a given (Overton, 1978); thus, the task of the scientist is to describe, explain, and optimize the parameters and trajectory of processes (i.e., variables that show time-related changes in their quantity and/or quality).

The second assumption of contextualism is thus raised. It stresses the interrelation of all levels of analysis. Because phenomena are not seen as static, but rather as change processes, and because any change process occurs within a similarly (i.e., constantly) changing world (of processes), any target change must be conceptualized in the context of the other changes within which it is embedded. Thus, change will constantly continue as a consequence of this embeddedness. In sum, these two focal points of contextualism appeared as key features of several conceptual appeals for this paradigm that occurred throughout the 1970s. We now consider a sample of these statements.

INSTANCES OF APPEALS FOR CONTEXTUALISM IN THE 1970s

Contextualism has been criticized as overintellectualized eclecticism and/or as an empirically or methodologically vacuous orientation (see Jenkins, 1974;

Lerner *et al.*, 1980; & Sarbin, 1977, for descriptions and reviews of these criticisms). Nevertheless, throughout the 1970s there were repeated calls for the use of contextually derived theory and research that were based precisely on the antithesis of these criticisms. Coming from a remarkably diverse array of intellectual traditions, these pleas suggested that contectualism offered both a conceptual framework for asking ecologically meaningful questions and suggested methodological strategies for doing new, and potentially more useful, empirical research. In this section, we illustrate the nature of these calls. In the following section, we consider whether there indeed is any empirical use for contextual thinking. A final section discusses the role of contextualism for the description, explanation, and optimization of human development.

As just indicated, calls for contextualism arose from psychologists associated with different intellectual traditions. For example, in 1974 James J. Jenkins rejected the mechanistic model he had used to guide his associationistic view of memory. He suggested that instead of this traditionally American approach to the study of memory, a contextual approach be adopted (Jenkins, 1974). He argued that "what memory is depends on context" (Jenkins, 1974, p. 789), and defended this view by presenting the results of several empirical studies which demonstrate that:

> What is remembered in a given situation depends on the physical and the psychological context in which the event was experienced, the knowledge and skills that the subject brings to the context, the situation in which we ask for evidence for remembering, and the relation of what the subject remembers to what the experimenter demands (Jenkins, 1974, p. 793).

Jenkins (1974, p. 787) noted that to adequately deal with all these sources of variation means that "being a psychologist is going to be much more difficult than we used to think it to be." In part, this difficulty arises because there is no one mode of analysis, or methodological strategy, that suggests itself as always useful for assessment of all the levels of analysis involved in memory processes at all historical moments. Thus, not only is methodological pluralism promoted from this contextual perspective, but too the criterion of usefulness must be employed when deciding if a particular methodological strategy is appropriate. That is, one must decide "What kind of an analysis of memory will be useful to you in the kinds of problems you are facing? What kind of events concern you?" (Jenkins, 1974, p. 794). In other words, Jenkins (1974, p. 794) believes that:

> The important thing is to pick the right kinds of events for your purposes. And it *is* true in this view that a whole theory of an experiment can be elaborated without contributing in an important way to the science because the situation is artificial and nonrepresentative in just the senses that determine its peculiar phenomena. In short, contextualism stresses relating one's laboratory problems to the ecologically valid problems of everyday life.

A similar call for a focus on the ecological validity of psychological research was independently put forth by Urie Bronfenbrenner in numerous publications throughout the 1970s. In these papers, Bronfenbrenner argued for an ecological approach to the study of human development. For example, in 1977 Bronfenbrenner noted that:

> much of contemporary developmental psychology is the science of *the strange behavior of children in strange situations with strange adults for the briefest possible periods of time* (Bronfenbrenner, 1977, p. 513).

Accordingly, he argued that only "Experiments created as real are real in their consequences" (Bronfenbrenner, 1977, p. 529), and, as did Jenkins (1974), called for methodological pluralism in order for progress to be made toward understanding human development in context.

To aid in the implementation of this perspective, Bronfenbrenner put forth a broad approach to research in human development that focused on the progressive accommodation through the life span between the developing human and the changing environments within which he or she actually lives and develops. These ecological milieus include not only the immediate settings containing the developing person but also the larger and more indirect social contexts, both formal and informal, within which the more immediate settings are embedded. The research methodologies that Bronfenbrenner suggests stress the use of rigorously designed experiments, both naturalistic and contrived, in order to appraise the changing relation between organism and environment. This developing relation is conceptualized in system terms, with the elements of the system involving several levels of analysis and the stress in the model being placed on the reciprocal relations among the levels of analysis.

This stress on the relations among the elements of a changing system composed of multiple levels of analysis, and not on the elements *per se,* is the key feature of the dialectic view of human development promoted by Klaus Riegel throughout the 1970s. As noted earlier in this paper, Riegel's (1975, 1976a, 1976b) view of human development involved the analysis of crises generated by asynchronies occurring along four dimensions of developmental progression: the inner-biological, the individual psychological, the cultural-sociological, and the outer-physical.

Riegel's conceptualization of the dialectical nature of human development placed emphasis on the study of the temporal order of concrete events brought about by conflicts and contradictions among these dimensions. He suggested that the dialogue be used as a methodological strategy to analyze event sequences and contradictions. As dialogues represent short-term changes, however, Riegel suggested that they be supplemented by studies of long-term changes across the life-span of individuals and the history of their society. Accordingly, he forwarded a dialectical psychology "manifesto" which, among other things, called for: (1) a commitment to the study of actions and changes, involving both short-term, situational alterations and long-term individual and cultural developments; and (2) a focus on the relations among all inner and outer dialectical changes such that the study of the activities of the individual, who invariably is at the intersection of these interactions, is the study of the relations of relations.

Riegel's (1976a) focus on dialogical methodology, and his stress on the reciprocal relation of the individual to his or her social context over the course of the sequence of events that comprise the person's life-span, are not dissimilar to the ideas presented by Sarbin (1977) in his call for contextualism as a world view for modern psychology.

Akin to Riegel's (1976a) dialogue is Sarbin's (1977) dramaturgical model, a technique which, through use of the notion of emplotment, attempts to capture the sequence of reciprocal events between individuals and their changing social contexts. Sarbin (1977) applies his contextualist model to the analysis of data sets pertinent to the genesis of schizophrenia, to the nature of hypnosis, and to the characteristics of imagination, in order to illustrate the integrative utility of contextually derived ideas. His presentation serves to illustrate that contextual ideas can be useful in understanding an array of psychological processes, ranging from those associated with cognition and affect to those traditionally labeled as personality and social ones; moreover, Sarbin stresses that the interrelation among processes can not only be integrated by contextual thinking but, in fact, needs to be appreciated if both adaptive and nonadaptive outcomes of person-context relations are to be understood. For example, Sarbin suggests that in the understanding of the bases of schizophrenia the contextualist will, as opposed to the mechanist, take:

> as his unit, not schizophrenia, not improper conduct, not the rules of society, but as much of the total context as he can assimilate. His minimal unit of study would be the man who acted as if he believed he could travel unaided through space *and* the person or persons who passed judgment on such claims (Sarbin, 1977, p. 25).

Thus, as in Riegel's (1976a) model of crises being generated by developmental asynchronies, Sarbin (1977) searches for the bases of adaptive and maladaptive functioning not within the realm of personological functioning, but rather within the domain of the conflicts and crises created by the degrees of "goodness of fit" (Thomas & Chess, 1977) a person experiences in his or her relations with the social context. Sarbin too sees the relevance of his ideas to those forwarded in other calls for contextualist thinking. In fact, he sees Jenkins (1974), as well as Cronbach (1975) and Gergen (1973), as making consonant appeals.

Indeed, not only are these latter two papers other instances of appeal for contextualism in the 1970s, but other prominent examples may be cited. The *American Psychologist* is the journal of the American Psychological Association designed to publish articles of current and broad interest to psychologists. The already-discussed papers by Jenkins (1974), Bronfenbrenner (1977), and Riegel (1977) were published in the *American Psychologist,* and in the last two years of the 1970s two additional papers appeared in the *American Psychologist* that, in different ways, made an appeal for contextualism. Petrinovich (1979) promoted an idea drawn from Egon Brunswik's notion of ecological validity, labeled by Petrinovich "probabilistic functionalism," which called for an array of methodological strategies not dissimilar in intent to those suggested in calls for methodological pluralism forwarded by contextual thinkers such as Bronfenbrenner (1977) and Jenkins (1974), among others (e.g., Lerner *et al.*, 1980). Most interestingly, Bandura (1978) reconceptualized his social learning theory as involving causal processes that are based on reciprocal determinism. Indeed, consistent with the two key contextual themes of constant change and embeddedness, Bandura asserted that, "from this perspective, psychological functioning involves a continuous reciprocal interaction between behavioral, cognitive, and environmental influences" (Bandura, 1978, p. 344).

We need not here expand on these last two instances of the promotion of contextural thinking to see that there was indeed a set of scholars, previously associated with a diverse array of conceptual orientations within psychology, who saw in the 1970s various theoretical, methodological, and empirical uses for contextually derived ideas. Turning now to two specific examples of research literatures that were advanced in the 1970s in new ways by such ideas, we will be able to illustrate whether and how these converging appeals have substance. In addition, these examples will allow us to draw conclusions not only about the future role of contextualism for theory, research, and method, but also for the devising of new strategies for the optimization of human development.

EMPIRICAL APPLICATIONS OF CONTEXTUALLY DERIVED CONCEPTS: TWO ILLUSTRATIONS

Although the appeal of contextualism in the 1970s has been seen to extend well beyond those developmental psychologists concerned with promoting a life-span view of human development, several research literatures within life-span developmental psychology present exemplars of the use of contextually derived ideas. Accordingly, to illustrate the past and future substantive contributions of contextualism for research we will draw on one of these literatures. We will discuss memory development during the adult and aged years. In addition, however, it will be useful to illustrate the substantive use of contextualism by reference to a research literature not only not derived from life-span developmental psychology, but further, not primarily associated with psychology at all. Thus, we will consider how contextual thinking reorders and extends our understanding of individual differences in temperament derived primarily from the research of the psychiatrists Thomas and Chess.

Memory Development during Adulthood and Aging

We have argued that during the 1970s psychologists became increasingly aware of the fact that theories and research are articulated within the context of world views. From such a perspective, descriptions of the memory process and the aging process will vary depending on the particular world view in which the theory and research are rooted (Reese, 1973, 1976; Riegel, 1977). Moreover, since the salience of various world views fluctuates over historical time, it follows that descriptions of the memory process and the aging process will also fluctuate over historical time. During the previous two decades, for example, both memory theory and aging prototheory have been largely derived from the organismic world view (Craik, 1977; Reese, 1973, 1976), whereas attention to models consistent with the contextual world view appear to be emerging during recent years (Baltes, Reese, & Lipsitt, 1980; Hultsch & Pentz, 1980; Meacham, 1977, Riegel, 1977).

Organismic Perspectives on Memory and Aging

In the case of memory theory, the assumptions of organicism are reflected in various multistore models of memory (Reese, 1973, 1976). These models postulate that material is transferred from one storage structure to another until a permanent memory trace is established. Typically, three levels of stores—sensory stores, a short-term store, and a long-term store—are identified (Atkinson & Schiffrin, 1968; Murdock, 1967). These stores are specifically defined by different encoding, storage, and retrieval characteristics (Craik & Lockhart, 1972). For example, coding in the sensory, short-term, and long-term stores is characterized as involving a literal copy, phonemic features, and semantic features, respectively; storage capacity is characterized as large, small, and unlimited, respectively; and retrieval is characterized as involving a direct readout, items in consciousness, and search processes and retrieval cues, respectively.

In the case of aging prototheory, the assumptions of organicism are reflected in the extension of maturational growth models to aging (Labouvie-Vief, 1977; Reese, 1973). Within such models, development is seen as biologically based and is characterized by unidirectional, irreversible, and universal sequences directed toward some end state (Harris, 1957). When extended into adulthood, such maturational growth models focus on intrinsic aging processes that result in the biological degeneration of the aging individual. Behaviorally, this has led to an emphasis on either a "decrement" or "decrement with compensation" view of cognition (Labouvie-Vief, 1977; Schaie, 1973). In the former instance, the emphasis is on the irreversible nature of change, whereas in the latter instance this emphasis is modified by an examination of environmental conditions that may ameliorate the deterioration. In both instances, however, the focus is on intrinsic processes that reflect "true" aging.

These organismically based assumptions about memory and aging combine to yield certain research emphases. In particular, the focus is on specifying the extent of the age-related decrement in memory performance and its locus within the encoding, storage, and retrieval processes of the various stores.

Contextual Perspectives on Memory and Aging

Different world views yield different concerns about memory and aging, however. For example, emerging work within both memory theory and aging prototheory is being articulated within the contextual world view.

In the case of memory theory, these assumptions are reflected in various "contextual" approaches to remembering (Bartlett, 1932; Bransford, McCarrell, Franks, & Nitsch, 1977; Jenkins, 1974; Meacham, 1977). Contextually based approaches view memory as a byproduct of the transaction between the individual and the context. Thus, the focus is on the nature of the events the individual experiences. Memory depends on the analysis of the event that takes place at many levels reflecting multiple contexts of increasing scope. What is learned and remembered, then, depends on the total context of the event; for example, the physical, psychological (perceptual, linguistic, semantic, and schematic), social, and cultural context in which the event was experienced, the context in which we ask for evidence of remembering, and so forth. Because of the emphasis on memory as a byproduct, remembering is not seen as an isolated process. Rather, the emphasis is on the interface of the various perceptual, inferential, linguistic, problem solving, personality, social, and cultural processes that contribute to the construction and reconstruction of events. Finally, because of the emphasis on a continuing transaction between the individual and the context, contextually based models do not assume the retrieval of a permanent memory trace. Rather, remembering is a reconstruction of past events. This depends in large measure on the degree to which the material has been articulated with past experience during acquisition. In addition, memory also depends on events occurring following acquisition. Thus, the individual continually constructs and reconstructs events as the context changes. The concepts of encoding, storage, and retrieval are fundamentally inconsistent with this view since they imply recovery of a static memory trace.

In the case of aging prototheory, the assumptions of contextualism are reflected in various life-span or dialectical views (Baltes & Willis, 1977; Riegel, 1976a). These views emphasize multidirectional change and multiple influences in development. That is, change during adulthood is assumed to involve different directions (incremental, decremental) and different forms (linear, curvilinear). As a result, interindividual differences tend to increase over the life span as intraindividual change functions become increasingly divergent. This concept departs sharply from the maturationally based view because it suggests a significant potential for growth during adulthood. Further, a contextually based perspective calls attention to multiple sources of influences on development, reflecting both biological and environmental antecedents at both the individual and cultural level. Finally, these sources are assumed to interact reciprocally with one another.

These contextually based assumptions about memory and aging combine to yield different research emphases. In particular, the focus is on growth as well as decline since multiple patterns of change are expected, depending on the context. Further, the focus is on the changing activities of perceiving, comprehending, and remembering rather than on the encoding, storage, and retrieval of a permanent memory trace.

Remembering in Adulthood

In sum, both memory theory and aging prototheory appear to be in conceptual transition toward a contextual world view. This convergence of trends suggests that the examination of adult learning and memory within a contextual framework is likely to be particularly productive. What does this convergence suggest about remembering in adulthood?

First, a contextual perspective draws our attention away from accuracy and toward meaning. Although remembering may be accurate, a contextual perspective emphasizes that the knowledge one gains from experience goes beyond what is shown in accurate recall or recognition. Memory for an event is likely to be far more complete and longer lasting than memory for the "surface" characteristics of the event (Bartlett, 1932; Cofer, 1977; Dooling & Christiaansen, 1977; Frederiksen, 1975; Kintsch, 1976). Obviously, accuracy is important. However, we have probably placed too much emphasis on it.

The potential contrast between meaning and accuracy in adult memory is illustrated by Walsh and Baldwin's (1977) recent study. These investigators presented Bransford and Franks' (1971) linguistic abstraction task to younger and older adults (average age 18.7 years and 67.5 years). Their general findings with the sample replicated those of Bransford and Franks. More interestingly, Walsh and Baldwin (1977) found no age differences in the integration of semantic information. A free-recall task administered to the same subjects, however, showed significant age differences with the older adults recalling only about half as much as the younger adults.

The emphasis on meaning is particularly important from an adult developmental perspective because of its relationship to ecological validity (Bronfenbrenner, 1977; Schaie, 1978). That many of our traditional learning and memory tasks lack meaning for older adults is apparent to anyone who has observed their behavior in the laboratory situation. The incidence of older adults' questions concerning the relevance of such tasks and refusals to continue participation is striking compared to that of younger adults. Such behaviors may reflect something more basic than general anxiety or uncooperativeness on the part of older adults. Thus, Schaie (1977) has suggested that there may be qualitatively different stages of adult cognitive functioning with acquisitive processes dominating in early adulthood and reintegrative processes dominating in later adulthood. These latter processes emphasize the meaningfulness of cognitive demands. To the extent that these observations are useful, it is important to modify our strategies for examining remembering in adulthood—including the development of a taxonomy of age-and-cohort specific life situations in which remembering occurs (Schaie, 1978; Siegel, 1977).

Second, a contextual perspective draws our attention away from narrow contexts and toward broad contexts. From a contextual perspective, acquisition involves the articulation of events with past experience. This differentiation and integration takes place at many levels reflecting wider and wider contexts including the physical, linguistic, semantic, schematic, functional, social, and cultural. The role of some of the narrower contexts of this hierarchy on adult remembering have received recent attention—particularly the contrast between the physical and semantic contexts of the input (Craik, 1977; Eysenck, 1974; Zelinsky, Walsh & Thompson, 1978). We have, however, ignored the higher levels almost entirely.

For example, little attention has been directed toward the role of world knowledge. From a contextual perspective, differences in such knowledge will have a significant effect on the comprehension and reconstruction of events (Dooling & Christiaansen, 1977; Kintsch, 1976). That different age or cohort groups will differ in such knowledge is obvious from the fact that they have experienced different individual cultural and historical events (Baltes, Cornelius, & Nesselroade, 1977; Hultsch & Plemons, 1979). The importance of these differences is probably crudely indexed by age changes and cohort differences in crystallized intelligence (Horn, 1970; Baltes & Labouvie, 1973; Schaie & Labouvie-Vief, 1974). For example, Gardner and Monge (1977) have recently demonstrated significant age/cohort-related patterns of knowledge in such domains as transportation, disease, slang, finance, religion, fashion, art, hobbies, sports, and current affairs.

Similarly, little attention has been paid to the relationship of acquisition and remembering to other activities—for example, intellectual, personality, and social processes. Works examining such multiprocess relationships in adulthood are almost nonexistent (Botwinick & Storandt, 1974; Fozard & Costa, 1977; Horn, 1978; Hultsch, Nesselroade & Plemons, 1976). What little work is available, although derived from noncontextual frameworks, suggests that such relationships may be of importance. For example, Hultsch, Nesselroade, and Plemons (1976) found age/cohort-related differences in patterns of ability-performance relationships on a free-recall task. Memory abilities were more predictive of performance for older adults than for younger adults, whereas the reverse is true for fluency abilities. Similarly, Fozard and Costa (1977) found age differences in patterns of personality-performance relationships on tasks measuring the speed of retrieval from primary, secondary, and tertiary memory.

Finally, we have paid little attention to the individual's goals. Remembering is a means to an end as well as an end in itself (Meacham, 1972, 1977). Further, as noted previously, it is likely that the goals of remembering change markedly over the life span. For example, there is some evidence to suggest the importance of remembering in constructing personality in late life (Butler, 1963; Meacham, 1977). Several studies have reported a relationship between reminiscence and various indices of adjustment (Boylin, Gordon & Nehrke, 1976; Costa & Kastenbaum, 1967; Havighurst & Glasser, 1972). Such a perspective reinforces the importance of memory in reconstructing the individual and the social context (Meacham, 1977).

Third, a contextual perspective draws our attention away from an emphasis solely on age-graded sources of influence. Baltes, Cornelius, and Nesselroade (1979), for instance, specify three major influence patterns that control development: individual age-graded influences, evolutionary history-graded influences, and nonnormative influences. The first two of these covary with time. On the other hand, individual age-graded influences are highly correlated with chronological age while, on the other hand, evolutionary history-graded influences are highly correlated with biocultural history. Influences from both systems may include biological as well as environmental variables (e.g., age changes in psychomotor speed or social interaction;

historical changes in the gene pool or the educational system). The third system—nonnormative influences—is not directly indexed by time since they do not occur universally for all people. In addition, when they do occur, they are likely to differ in terms of their clustering, timing, and duration. Nonnormative influences consist of what may be labeled life events (e.g., illness, divorce, promotion).

To date, our research on adult memory has focused almost entirely on individual age-graded sources of variance. Undeniably, performance differences between adults of different ages do occur—and typically, these reveal poorer performance on the part of older adults. However, in many respects, our research on adult memory has been both methodologically and conceptually inadequate with respect to the issue of multiple sources of influence. Methodologically, there is little research available on adult learning and memory that would allow the examination of change—the fundamental concern of developmentalists—and still less which uses any form of sequential analyses to partially distinguish age-graded and history-graded sources of influence (Arenberg & Robertson-Tchabo, 1977; Gilbert, 1936, 1973). More data are required in order to examine these issues. Conceptually, little attention has been directed toward the potential impact of history-graded or nonnormative sources of influence as antecedents of age-related changes or differences in memory performance. This does not mean that ontogenetic age-graded sources of influence are unimportant. For example, it is likely that the slowing of the central nervous system plays a significant role in some aspects of adult memory. However, other non-age-related sources of influence, such as historical changes in educational patterns, are also likely to be critical. From a contextual perspective, such historical variables are just as "developmentally relevant" as changes in speed. Further, our own hunch is that the more we emphasize meaning rather than accuracy and the role of broad rather than narrow contexts in remembering, the more significant evolutionary history-graded and nonnormative influences will become. In sum, a contextual orientation significantly revises the study of memory development during the adult and aged years. Contextual thinking also revises the study of temperament.

Temperament and Psychosocial Adaption

In the early decades of this century the study of temperament was associated with a constitutional point of view (e.g., Kretschmer, 1925) linked to an organicist paradigm. During this century's middle decades other instances of constitutionalism, notably that of Sheldon (1940, 1942), were forwarded along with, alternative, mechanistically derived psychometric approaches to the conceptualization and study of temperament (e.g., Cattell, 1950; Guilford, 1959; Guilford & Zimmerman, 1949). During the 1960s, and continuing through this writing, interest in the social sciences and medicine in temperament derives primarily from the work of Thomas and Chess (e.g., Thomas & Chess, 1977; Thomas, Chess, & Birch, 1970). To Thomas and

Chess (1977) temperament is conceived of as the stylistic component of behavior, i.e., how an organism does whatever it does; their work focuses on how individual differences, along several dimensions of temperament uncovered in their research, covary with adaptive and maladaptive psychosocial functioning. For example, all children engage in eating, sleeping, and toileting behaviors. While attention to the absence or presence of such contents of the behavior repertoire would not easily differentiate among children, focus on whether these behaviors occur with regularity (i.e., rhythmically or predictably), with a lot or a little motor activity, intensity, or vigor, or whether there is a negative, positive, or neutral mood associated with the behaviors, might serve to differentiate among adaptive versus nonadaptive children.

In fact, particular constellations of temperamental attributes have been found to place children at risk for the development of problem behaviors. Results from the Thomas and Chess New York Longitudinal Study (NYLS; Thomas, Chess & Birch, 1968; Thomas *et al.,* 1963) indicate that children having a temperamental repertoire labeled as difficult (e.g., low rhythmicity, negative mood, and high intensity) have higher incidences of behavioral or emotional disorders than do children having repertoires labeled as easy (e.g., high rhythmicity, positive moood, and moderate intensity) or slow-to-warm-up (e.g., initial withdrawal behaviors and negative mood). For example, in one report (Thomas *et al.,* 1970) 42% of the total number of children in the NYLS sample eventually developed behavior problems severe enough to call for psychiatric attention. About 70% of the difficult children developed such problems, while only 18% of the easy children did so; and the percentage of slow-to-warm-up children who developed such problems was between that of these other two groups. Although the NYLS data are derived from an essentially white, middle-class sample, data from other samples, collected by the Thomas group (e.g., Korn, Chess & Fernandez, 1978; Thomas & Chess, 1977) and others (Sameroff, 1978), confirm the linkages between differential temperamental repertoires and contrasting psychosocial developments.

Recently, Thomas and Chess (1981) have emphasized that the metatheoretical model from which their conceptualization of temperament is derived is a contextual one. Thus, they suggest that the impact of temperament for adaptive functioning lies not in the possession of a particular repertoire per se. Rather, the impact lies in whether a particular repertoire provides a "goodness of fit" with the individual demands of a specific context (Thomas & Chess, 1977). Analogous ideas about organism-environment congruence have been forwarded by Riegel (1975, 1976a, 1976b) and by Sarbin (1977).

Using this perspective it has been suggested (Korn, 1978) that it may be that white, middle-class social contexts have fairly generalizable views about desirable behavioral styles for children. If so, then a child with a repertoire that has been labeled as difficult is only "at risk" insofar as his or her arrhythmicity, negative mood, and high intensity reactions are not congruent with such demands. However, in another context, having alternative appraisals of such attributes, the "at risk" status would change. In other words, a contextual perspective suggests that labeling a child as difficult is a misleading designation, since it places the locus of the problem *in* the child. Only when there is a mismatch *between* a particular temperament and a particular context may difficulty arise. As such, the issue in temperament research is whether a particular set of person attributes are congruent or incongruent with the demands of a specific context. One should ask what person characteristics in interaction with what environmental characteristics lead to what outcomes.

This contextual perspective serves to reorder extant data sets. Korn (1978) and Gannon (1978) have presented data indicating that in lower-class Puerto Rican settings the attributes that define the difficult child syndrome are not only not undesirable among parents but, in turn, may be highly regarded. Thus, as opposed to white, middle-class children, a child in the Puerto Rican sample would not have a repertoire incongruent with the demands of the context. Accordingly, among such children such a temperament should not be as highly associated with nonadaptive developments as is the case in white, middle-class samples. Indeed, this is precisely the findings reported by Korn (1978) and Gannon (1978). Similarly, Super (1978) finds that in rural Kenyan mothers, who stay in almost complete physical contact with their infants for the first few years of the child's life, components of the difficult child syndrome are not as problematic for the mother-child relation as is the case when a difficult child is born to an urban Kenyan mother. These urban mothers are not in as constant physical contact with their children. Thus, for example, infant irregularity or unpredictability regarding hunger creates more problems for a breast-feeding urban mother than for a breast-feeding rural one.

The contextual conceptualization of temperament may also extend temperament research. While previous research has typically looked at the goodness of fit within one context, current contextually derived temperament studies have considered how temperament provides goodness of fits across contexts (e.g., J. Lerner, 1980). The notion here is that those children showing the most plastic repertoire, i.e., a repertoire that alters in relation to the changing demands extant in diverse contexts, will be those children who are most psychosocially adaptive.

Finally, past and current contextually based research on temperament provides new ideas for intervention. Armed with the knowledge of what person characteristics in relation to what environmental characteristics lead to what development outcomes, an interventionist could focus on enhancing optimal relations both within and among contexts. These interventions may involve work at multiple levels of analysis, and may often be those levels not directly related to the target relation per se. For example, attempts to alter attitudes and values regarding particular temperamental characteristics may prevent temperament-context mismatches, and such interventions may be directed at the social network, community, or educational institution. We have seen Bronfenbrenner (1977) make similar suggestions.

In sum, contextually derived ideas not only alter the description and explanation of human developmental processes, but too, insofar as intervention is concerned, they suggest that a pluralism of targets, timings, and mechanisms may be involved in attempts to optimize people's functioning.

This assertion leads us into some concluding comments about the future of contextualism.

CONCLUSIONS: THE FUTURE USE OF CONTEXTUALISM FOR THEORY, RESEARCH, AND INTERVENTION

If the interest in and the use of contextually derived ideas that burgeoned in the 1970s continues into the 1980s, we believe there will be significant changes in the nature of developmental psychology, in particular, and in the entire science and practice of psychology, in general. These changes will come about primarily as a consequence of the impact contextualism will have on theory and research.

Contextualism offers an alternative developmental theory that focuses on constant change and embeddedness. If these emphases influence psychology in general then their effect will be to make other substantive areas within psychology developmental in focus. In addition, the idea of embeddedness leads to an interest in the multiple levels or contexts of development, and such concern alters the focus in developmental research from internal to external validity (Hultsch & Hickey, 1978). In addition, the concern with the relations among levels, promoted by the notion of embeddedness, leads to an interest in multidisciplinary research. Together, these contextual orientations may coalesce in a movement away from psychological models of psychological functioning.

In addition, contextualism's stress on the potential significance of all events in the context may be associated with major alterations in the methodological repertoire of human developmentalists. The perspective of the subject as well as the researcher may be important sources of information about variation in behavior change processes. Such recognition suggests the use of less mechanistic, but often richer, research tools. For example, phenomenological and biographical data collection strategies may be used more frequently. Such inclusions in our methodological repertoire have the interesting side effect of humanizing our study of human development. The perspective of the developing person becomes an integral part of all complete data sets.

The notions of constant change and of embeddedness also have implications for intervention. The focus on constant embedded change suggests the potential plasticity of human functioning, and implies that multiple levels of analysis and contexts are involved in any behavior development. These recognitions may revise the questions asked in human development intervention. From a contextual perspective, one should ask what techniques are best suited for what behaviors of what people at what points in their lives. To paraphrase Jenkins (1974), contextualism thus makes the job of intervening even more difficult than we now know it to be.

But, rather than becoming pessimistic about the difficulty of optimizing human functioning, it is our view that an appropriate understanding of contextualism provides an optimistic view of human functioning and human potential. Given the willingness to combine our skills with those whose intervention abilities lie at other levels of analysis and/or with other mechanisms for change, we can assume that any target behavior is amenable to at least some modification. The contextual character of human development thus asserts the basic plasticity of human behavior. It also gives us all the ability to contribute effectively to the historical events that surround us, and in so doing become producers of the future history of more optimal human development.

REFERENCES

ANASTASI, A. 1958. Heredity, environment, and the question "how?" Psychological Review **65:** 197–208.

ARENBERG, D. & E. A. ROBERTSON-TCHABO. 1977. Learning and aging. *In* Handbook of the Psychology of Aging. J. E. Birren & K. W. Schaie, Eds. Van Nostrand Reinhold. New York.

ATKINSON, R. C., & R. M. SHIFFRIN. 1968. Human memory: A proposed system and its control processes. *In* The Psychology of Learning and Motivation, Vol. 2. K. W. Spence & J. T. Spence, Eds. Academic Press. New York.

BALTES, P. B., Ed. 1978. Life-span Development and Behavior, Vol. 1. Academic Press. New York.

BALTES, P. B. 1979a. Life-span developmental psychology: Some converging observations on history and theory. *In* Life-span Development and Behavior, Vol. 2. P. B. Baltes & O. G. Brim, Jr., Eds. Academic Press. New York.

BALTES, P. B. 1979b. On the potential and limits of child development: Life-span developmental perspectives. Newsletter of the Society for Research in Child Development, (Summer): 1–4.

BALTES, P. B. & M. M. BALTES. 1979c. Plasticity and variability in psychological aging: Methodological and theoretical issues. *In* Department of Gerontology, Institute of Neuropsychopharmacology, Free University of Berlin: Symposium contribution to appear in "Methodological considerations in determining the effects of aging on the CNS." Berlin, West Germany: July 5–7, 1979.

BALTES, P. B., M. M. BALTES & G. REINERT. 1970. The relationship between time of measurement and age in cognitive development of children: An application of cross-sectional sequences. Human Development **13:** 258–268.

BALTES, P. B. & O. BRIM, Eds. 1979. Life-span Behavior and Development, Vol. 2. Academic Press. New York.

BALTES, P. B., S. W. CORNELIUS & J. R. NESSELROADE. 1977. Cohort effects in behavioral development: Theoretical and methodological perspectives. *In* Minnesota Symposia on Child Psychology, Vol. II. W. A. Collins, Ed. Thomas Crowell. New York.

BALTES, P. B., S. W. CORNELIUS & J. R. NESSELROADE. 1979b. Cohort effects in developmental psychology. *In* Longitudinal Research in the Study of Behavior and Development. J. R. Nesselroade & P. B. Baltes, Eds. Academic Press. New York.

BALTES, P. B. & J. R. NESSELROADE. 1973. The developmental analysis of individual differences on multiple measures. *In* Life-span Developmental Psychology: Methodological Issues. J. R. Nesselroade & H. W. Reese, Eds. Academic Press. New York.

BALTES, P. B. & G. V. LABOUVIE. 1973. Adult development of intellectual performance: Description, explanation, and modification. *In* The Psychology of Adult

Development and Aging. C. Eisdorfer & M. P. Lawton, Eds. American Psychological Association. Washington, D.C.

BALTES, P. B., H. W. REESE & J. R. NESSELROADE. 1977. Life-span Developmental Psychology: Introduction to Research Methods. Brooks-Cole. Monterey, CA.

BALTES, P. B., H. W. REESE & L. P. LIPSITT. 1980. Life-span developmental psychology. Annual Review of Psychology **31:** 65–110.

BALTES, P. B. & K. W. SCHAIE, Eds. 1973. Life-span Developmental Psychology: Personality and Socialization. Academic Press. New York.

BALTES, P. B. & K. W. SCHAIE. 1974. The myth of the twilight years. Psychology Today **7:** 35–40.

BALTES, P. B. & K. W. SCHAIE. 1976. On the plasticity of intelligence in adulthood and old age: Where Horn and Donaldson fail. American Psychologist **31:** 720–725.

BALTES, P. B. & S. L. WILLIS. 1977. Toward psychological theories of aging and development. *In* Handbook of the Psychology of Aging. J. E. Birren & K. W. Schaie, Eds. Van Nostrand Reinhold. New York.

BANDURA, A. 1978. The self system in reciprocal determinism. American Psychologist **33:** 344–358.

BARTLETT, F. C. 1932. Remembering. Cambridge University Press. Cambridge.

BOYLIN, W., S. K. GORDON & M. F. NEHRKE. 1976. Reminiscing and ego integrity in institutionalized elderly males. Gerontologist **16:** 118–124.

BOTWINICK, J. & M. STORANDT. 1974. Memory, Related Functions, and Age. C. C. Thomas. Springfield, IL.

BRANSFORD, J. D. & J. J. FRANKS. 1971. The abstraction of linguistic ideas. Cognitive Psychology **2:** 331–350.

BRANSFORD, J. D., N. S. MCCARRELL, J. J. FRANKS & K. E. NITSCH. 1977. Toward unexplaining memory. *In* Perceiving, Acting, and Knowing: Toward an Ecological Psychology. R. Shaw & J. D. Bransford, Eds. Erlbaum. Hillsdale, NJ.

BRONFENBRENNER, U. 1963. Development theory in transition. *In* Child Psychology. Sixty-second Yearbook of the National Society for the Study of Education, Part I. H. W. Stevenson, Ed. University of Chicago Press. Chicago.

BRONFENBRENNER, U. 1977. Toward an experimental ecology of human development. American Psychologist **32:** 513–531.

BURGESS, R. L. & T. L. HUSTON, Eds. 1979. Social Exchange in Developing Relationships. Academic Press. New York.

BUTLER, R. 1963. The life review: An interpretation of reminiscence in the aged. Psychiatry **26:** 65–76.

CAPLAN, A. L., Ed. 1978. The Sociobiology Debate. Harper & Row. New York.

CARLSON, R. 1972. Understanding women: Implications for personality theory and research. Journal of Social Issues **28:** 17–32.

CATTELL, R. B. 1950. Personality: A Systematic, Theoretical, and Factual Study. McGraw-Hill. New York.

COFER, C. N. 1977. On the constructive theory of memory. *In* The Structuring of Experience. F. Weizman & I. C. Uzgiris, Eds. Plenum. New York.

COSTA, P. T. & R. KASTENBAUM. 1967. Some aspects of memories and ambitions in centenarians. Journal of Genetic Psychology **110:** 3–16.

CRAIK, F. I. M. 1977. Age differences in human memory. *In* Handbook of the Psychology of Aging. J. E. Birren & K. W. Schaie, Eds. Van Nostrand Reinhold. New York.

CRAIK, F. I. M. & R. S. LOCKHART. 1972. Levels of processing: A framework for memory research. Journal of Verbal Learning and Verbal Behavior **11:** 671–684.

CRONBACH, L. J. 1975. Beyond the two disciplines of scientific psychology. American Psychologist **30:** 116–127.

DATAN, N. & L. H. GINSBERG, Eds. 1975. Life-span Developmental Psychology: Normative Life Crises. Academic Press. New York.

DATAN, N. & H. W. REESE, Eds. 1977. Life-span Developmental Psychology: Dialectical Perspective on Experimental Psychology. Academic Press. New York.

DOOLING, J. D. & R. E. CHRISTIAANSEN. 1977. Levels of encoding and retention of prose. In The Psychology of Learning and Memory, Vol. 11. G. H. Bower, Ed., Academic Press. New York.

ELDER, G. H., Jr. 1974. Children of the Great Depression. University of Chicago Press. Chicago.

ELDER, G. H., Jr. 1979. Historical change in life patterns and personality. In Life-span Development and Behavior. Vol. 2. P. B. Baltes & O. G. Brim, Eds. Academic Press. New York.

EYSENCK, H. J. 1971. The IQ Argument: Race, Intelligence and Education. Library Press. New York.

EYSENCK, M. W. 1974. Age differences in incidental learning. Development Psychology 10: 936–941.

FELDMAN, M. W. & R. C. LEWONTIN. 1975. The heritability hang-up. Science 190: 1163–1168.

FOZARD, J. L. & P. T. COSTA. 1977. Age differences in memory and decision-making in relation to personality, abilities, and endocrine function: Implications for clinical practice and health planning policy. Paper presented at the Conference on Aging and Social Policy, Vichy, France, May, 1977.

FREDERIKSEN, C. H. 1975. Effects of context-induced processing operations on semantic information acquired from discourse. Cognitive Psychology 7: 139–166.

GANNON, S. 1978. Behavioral problems and temperament in middle-class and Puerto Rican five year old boys. Unpublished master's thesis, Hunter College, CUNY, New York, NY.

GARDNER, E. F. & R. H. MONGE. 1977. Adult age differences in cognitive abilities and educational background. Experimental Aging Research 3: 337–383.

GERGEN, K. J. 1973. Social psychology and history. Journal of Personality and Social Psychology 26: 309–320.

GILBERT, J. G. 1936. Mental efficiency in senescence. Archives of Psychology. 27: 188.

GILBERT, J. G. 1973. Thirty-five year follow-up study of intellectual functioning. Journal of Gerontology. 28: 68–72.

GUILFORD, J. P. 1959. Personality. McGraw-Hill. New York.

GUILFORD, J. P. & W. S. ZIMMERMAN. 1949. The Guilford-Zimmerman Temperament Survey: Manual of Instructions and Interpretations. Sheridan Supply Co. Beverly Hills, CA.

GOULET, L. R. & P. B. BALTES, Eds. 1970. Life-span development psychology: Research and theory. Academic Press. New York.

HARTUP, W. W. 1978. Perspectives on child and family interaction: Past, present, and future. In R. M. Lerner & B. G. Spanier, Eds. Child Influences on Marital and Family Interaction: A Life-span Perspective. Academic Press. New York.

HAVIGHURST, R. J. & R. GLASSER. 1972. An exploratory study of reminiscence. Journal of Gerontology. 27: 245–253.

HEBB, D. O. 1949. The Organization of Behavior. Wiley. New York.

HEBB, D. O. 1958. A Textbook of Psychology. W. B. Saunders Company. Philadelphia.

HEBB, D. O. 1970. A return to Jensen and his social critics. American Psychologist 25: 568.

HEMPEL, C. G. 1966. Philosophy of Natural Science. Prentice-Hall. Englewood Cliffs, NJ.

HERRNSTEIN, R. J. 1971. I.Q. Atlantic Monthly. **228:** 43–64.

HILL, R. & P. MATTESSICH. 1979. Family development theory and life-span development. *In* Life-span Development and Behavior, Vol. 2. P. B. Baltes & O. G. Brim. Jr. Eds. Academic Press. New York.

HIRSCH, J. 1970. Behavior-genetic analysis and its biological consequences. Seminars in Psychiatry. **2:** 89–105.

HORN, J. L. 1970. Organization of data on life-span development of human abilities. *In* Life-span Developmental Psychology: Theory and Research. L. R. Goulet & P. B. Baltes, Eds. Academic Press. New York.

HORN, J. L. 1978. Human ability systems. *In* Life-span Development and Behavior, Vol. 1. P. B. BALTES, Ed. Academic Press. New York.

HULTSCH, D. F. & T. HICKEY. 1978. External validity in the study of human development: Theoretical and methodological issues. Human Development **21:** 76–91.

HULTSCH, D. F. & C. A. PENTZ. 1980. Encoding, storage, and retrieval in adult memory: The role of model assumptions. *In* New Directions in Memory and Aging: Proceedings of the George A. Talland Memorial Conference. L. W. Poon, J. L. Fozard, L. S. Cermak, D. Arenberg & L. W. Thompson, Eds. Erlbaum. Hillsdale, NJ.

HULTSCH, D. F. & J. K. PLEMONS. 1979. Life events and lifespan development. *In* Lifespan Development and Behavior, Vol. 2. P. B. Baltes & O. G. Brim, Jr., Eds. Academic Press. New York.

HULTSCH, D. F., J. R. NESSELROADE & J. K. PLEMONS. 1976. Learning-ability relations in adulthood. Human Development **19:** 234–247.

JENKINS, J. J. 1974. Remember that old theory of memory? Well forget it. American Psychologist **29:** 785–795.

JENSEN, A. R. 1969. How much can we boost IQ and scholastic achievement? Harvard Educational Review **39:** 1–123.

JENSEN, A. R. 1973. Educability and Group Differences. Harper and Row. New York.

KAMIN, L. J. 1974. The science and politics of IQ. Erlbaum. Potomac, MD.

KAUFMANN, H. 1968. Introduction to the Study of Human Behavior. Saunders. Philadelphia.

KINTSCH, W. 1976. Memory for prose. *In* The Structure of Human Memory. C. N. Cofer, Ed. Freeman. San Francisco.

KORN, S. J. 1978. Temperament, vulnerability and behavior disorder. Paper presented at the Louisville Temperament Conference. Louisville, KY, September, 1978.

KORN, S. J., S. CHESS & P. FERNANDEZ. 1978. The impact of children's physical handicaps on marital quality and family interaction. *In* Child Influences on Marital and Family Interaction: A Life-span Perspective. R. M. Lerner, G. B. Spanier, Eds. Academic Press. New York.

KRETSCHMER, E. 1925. Physique and character. Harcourt. New York.

KUHN, D. 1978. Mechanisms of cognitive and social development: One psychology or two? Human Development **21:** 92–118.

KUHN, T. S. 1970. The Structure of Scientific Revolutions, Rev. edit. University of Chicago Press. Chicago.

LABOUVIE-VIEF, G. 1977. Adult cognitive development: In search of alternative interpretations. Merrill-Palmer Quarterly, **23:** 227–263.

LAYZER, D. 1974. Heritability analyses or IQ scores: Science or numerology. Science **183:** 1259–1266.

LEHRMAN, D. S. 1953. A critique of Konrad Lorenz's theory of instinctive behavior. Quarterly Review of Biology **28:** 337–363.

LEHRMAN, D. S. 1970. Semantic and conceptual issues in the nature-nurture problem. *In* Development and Evolution of Behavior: Essays in Memory of T. C. Schneirla. L. R. Aronson, E. Tobach, D. S. Lehrman & J. S. Rosenblatt, Eds. W. H. Freeman. San Francisco.

LERNER, J. V. 1980. The role of congruence between temperament and school demands in school children's academic performance, personal adjustment, and social relations. Unpublished Ph.D. dissertation. The Pennsylvania State University, University Park, PA.

LERNER, R. 1976. Concepts and Theories of Human Development. Reading, MA. Addison-Wesley.

LERNER, R. M. 1978. Nature, nurture, and dynamic interactionism. Human Development **21:** 1–20.

LERNER, R. M. 1979. A dynamic interactional concept of individual and social relationship development. *In* Social Exchange in Developing Relationships. R. L. Burgess & T. L. Huston, Eds. Academic Press. New York.

LERNER, R. M., E. A. SKINNER & G. T. SORELL. 1980. Methdological implications of contextual/dialectic theories of development. Human Development. **23:** 225–235.

LERNER, R. M. & G. B. SPANIER, Eds. 1978. Child Influences on Marital and Family Interaction: A Life-span Perspective. Academic Press. New York.

LEWONTIN, R. C. 1976. The fallacy of biological determinism. The Sciences **16:** 6–10.

LOEHLIN, J. C., G. LINDZEY & J. N. SPUHLER. 1975. Race Differences in Intelligence. W. H. Freeman. San Francisco.

LOOFT, W. R. 1972. The evolution of developmental psychology: A comparison of handbooks. Human Development **15:** 187–201.

LOOFT, W. R. 1973. Socialization and personality throughout the life-span. An examination of contemporary psychological aproaches. *In* Life-span Developmental Psychology: Personality and Socialization. P. B Baltes & K. W. Schaie, Eds. Academic Press. New York.

MEACHAM, J. A. 1972. The development of memory abilities in the individual and society. Human Development **15:** 205–228.

MEACHAM, J. A. 1977. A transactional model of remembering. *In* N. Datan & H. W. Reese, Eds. Life-span Developmental Psychology: Dialectical Perspectives on Experimental Research. Academic Press. New York.

MURDOCK, B. B. 1967. Recent developments in short-term memory. British Journal of Psychology. **58:** 421–433.

MUSSEN, P. H., Ed. 1970. Carmichael's manual of child psychology, 3rd edit. John Wiley. New York.

NESSELROADE, J. R. & P. B. BALTES. 1974. Adolescent personality development and historical change: 1970–1972. Monographs of the Society for Research in Child Development. **39:** (1, Serial No. 154).

NESSELROADE, J. R. & H. W. REESE, Eds. 1973. Life-span Developmental Psychology: Methodological Issues. Academic Press. New York.

OVERTON, W. F. 1973. On the assumptive base of the nature-nurture controversy: Additive versus interactive conceptions. Human Development. **16:** 74–89.

OVERTON, W. F. 1978. Klaus Riegel: Theoretical contribution to concepts of stability and change. Human Development. **21:** 360–363.

OVERTON, W. F. & H. W. REESE. 1973. Models of development: Methodological implications. *In* Life-span Developmental Psychology: Methodological Issues. J. R. Nesselroade & H. W. Reese, Eds. Academic Press. New York.

PEPPER, S. C. 1942. World Hypotheses: A Study in Evidence. University of California Press. Berkeley.

PETRINOVICH, L. 1979. Probabilistic functionalism: A conception of research method. American Psychologist. **34:** 373–390.

REESE, H. W. 1973. Models of memory and models of development. Human Development. **16:** 397–416.

REESE, H. W. 1976. Models of memory development. Human Development. **19:** 291–303.

REESE, H. W. & W. F. OVERTON. 1970. Models of development and theories of development. *In* Life-span Developmental Psychology: Research and Theory. L. R. Goulet & P. B. Baltes, Eds. Academic Press. New York.

RIEGEL, K. F. 1972. The influence of economic and political ideology upon the development of developmental psychology. Psychological Bulletin. **78:** 129–141.

RIEGEL, K. R. 1973. Developmental psychology and society: Some historical and ethical considerations. *In* Life-span Developmental Psychology: Methodological Issues. J. R. Nesselroade & H. W. Reese, Eds. Academic Press. New York.

RIEGEL, K. F. 1975. Toward a dialectical theory of development. Human Development. **18:** 50–64.

RIEGEL, K. F. 1976a. The dialectics of human development. American Psychologist. **31:** 689–700.

RIEGEL, K. F. 1976b. From traits and equilibrium toward developmental dialectics. *In* Nebraska Symposium on Motivation. W. Arnold, Ed. University of Nebraska Press. Lincoln.

RIEGEL, K. F. 1977. History of psychological gerontology. *In* Handbook of the Psychology of Aging. J. E. Birren & K. W. Schaie, Eds. Van Nostrand Reinhold. New York.

RILEY, M. W., Ed. 1979. Aging from birth to death. American Association for the Advancement of Science. Washington, D.C.

SAMEROFF, A. 1975. Transactional models in early social relations. Human Development. **18:** 65–79.

SAMEROFF, A. J. 1978. Differences in infant temperament in relation to maternal mental illness and race. Paper presented at the Louisville Temperament Conference, Louisville, KY, September, 1978.

SARBIN, T. R. 1977. Contextualism: A world view for modern psychology. *In* Nebraska Symposium on Motivation, 1976. J. K. Cole, Ed. University of Nebraska Press. Lincoln.

SCARR-SALAPATEK, S. 1971. Race, social class, and IQ. Science **174:** 1285–1295.

SCHAIE, K. W. 1973. Methodological problems in descriptive developmental research on adulthood and aging. *In* Life-span Developmental Psychology: Methodological Issues. J. R. Nesselroade & H. W. Reese, Eds. Academic Press. New York.

SCHAIE, K. W. 1977. Toward a stage theory of adult cognitive development. International Journal of Aging and Human Development. **8:** 129–128.

SCHAIE, K. W. 1978. External validity in the assessment of intellectual development in adulthood. Journal of Gerontology. **33:** 695–701.

SCHAIE, K. W. & G. LABOUVIE-VIEF. 1974. Generational versus ontogenetic components of change in adult cognitive behavior: A fourteen-year-cross-sequential study. Developmental Psychology. **10:** 305–320.

SCHAIE, K. W., G. V. LABOUVIE & B. V. BUECH. 1973. Generational and cohort-specific differences in adult cognitive functioning: A fourteen-year study of independent samples. Developmental Psychology. **9:** 151–166.

SCHNEIRLA, T. C. 1957. The concept of development in comparative psychology. *In* The Concept of Development. D. B. Harris, Ed. University of Minnesota Press. Minneapolis.

SEARS, R. R. 1975. Your ancients revisited: A history of child development. *In* Review of child development research, Vol. 5. E. M. Hetherington, Ed. University of Chicago Press. Chicago.

SHELDON, W. H. 1940. The Varieties of Human Physique. Harper & Row. New York.

SHELDON, W. H. 1942. The varieties of temperament. Harper & Row. New York.

SIEGEL, A. W. 1977. "Remembering" is alive and well (and even thriving) in empiricism. *In* Life-span Developmental Psychology: Dialectical Perspectives in Experimental Research. N. Datan & H. W. Reese, Eds. Academic Press. New York.

SUPER, C. 1978. Temperament and socialization among rural Kenyan infants. Paper presented at the Louisville Temperament Conference, Louisville, KY, September, 1978.

THOMAS, A. & S. CHESS. 1977. Temperament and development. Brunner/Mazel. New York.

THOMAS, A. & S. CHESS. 1981. The role of temperament in the contributions of individuals to their development. *In* Individuals as Producers of Their Development: A Life-span Perspective. R. M. Lerner & N. A. Busch-Rosnagel, Eds. Academic Press. New York.

THOMAS, A., S. CHESS & H. G. BIRCH. 1968. Temperament and Behavior Disorders in Children. New York University Press. New York.

THOMAS, A., S. CHESS & H. G. BIRCH. 1970. The origin of personality. Scientific American. **223:** 102–109.

THOMAS, A., S. CHESS, H. G. BIRCH, M. HERTZIG & S. J. KORN. 1963. Behavioral Individuality in Early Childhood. New York University Press. New York.

TOBACH, E., J. GIANUTSOS, H. R. TOPOFF & C. G. GROSS. 1974. The Four Horses: Racism, Sexism, Militarism, and Social Darwinism. Behavioral Publications. New York.

WALSH, D. A. & M. BALDWIN. 1977. Age differences in integrated semantic memory. Developmental Psychology. **13:** 509–514.

WILSON, E. O. 1975. Sociobiology: The New Synthesis. Harvard University Press. Cambridge, MA.

ZELINSKI, E. M., D. A. WALSH & L. W. THOMPSON. 1978. Orienting task effects on EDR and free recall in three age groups. Journal of Gerontology. **33:** 239–245.

Lebesgue's Measure Problem and Zermelo's Axiom of Choice: The Mathematical Effects of a Philosophical Dispute[a]

GREGORY H. MOORE

Department of Mathematics
University of Toronto
Toronto, Canada M5S 1A1

INTRODUCTION

During the early years of the twentieth century there arose an intriguing historical contradiction. This was not a contradiction within mathematics *per se,* such as was furnished during the same period by Russell's paradox, Burali-Forti's paradox, and the paradox of the largest cardinal. Rather, it was a contradiction between the expressed *philosophy* of a group of mathematicians and the type of *mathematics* that they developed. The contradiction was so fundamental that, if these mathematicians had understood it properly, their philosophical beliefs would have forced them to reject their life's work. At the very least, they would have had to recast it in a form that would destroy its usefulness to the mainstream of twentieth-century mathematics.

In order to understand this contradiction and its effects on the subsequent development of mathematics, we must first consider how Georg Cantor's set theory influenced French analysis late in the nineteenth century. In 1870 Cantor began to publish his findings on trigonometric series and, in particular, on what are now called their sets of uniqueness (*i.e.,* sets on which a real function can be represented by a unique trigonometric series). Little by little, this research led him to introduce his infinite ordinal and cardinal numbers as well as various concepts in point-set topology. Yet his ideas had a negligible impact on French mathematics until 1883 when Gösta Mittag-Leffler, editor of the new Swedish journal *Acta Mathematica,* printed French translations of several of Cantor's articles.[b] At the same time Cantor published in *Acta* a new but related paper [1883k], written in French.

The introduction of Cantor's ideas into France was largely due to two brothers, Jules and Paul Tannery. In the *Bulletin des sciences mathématiques et astronomiques,* whose audience consisted primarily of mathematicians, Jules Tannery [1884] favorably reviewed both the French translations of

---

[a]This paper was presented at a meeting held jointly by the Section of History, Philosophy and Ethical Issues of Science and Technology and the Section of Mathematics of The New York Academy of Sciences on March 25, 1981.
[b]The articles were Cantor, 1871, 1872, 1874, 1878, 1879, 1880, 1882, 1883, 1883a, and the corresponding translations were Cantor, 1883b–j. For a detailed and insightful study of Cantor's life and work, see Dauben, 1979.

0077–8923/83/0412–0129 $01.75/2 © 1983, NYAS

Cantor's work and his new paper. Indeed, Jules Tannery described Cantor's researches as constituting "a theory which, undoubtedly, is neither finished nor perfect but which is interesting from a philosophical viewpoint and which, from a mathematical perspective, is intimately connected to recent discoveries in the theory of [complex] functions . . ." [1884, p. 162]. In particular, Jules Tannery did not object to Cantor's infinite ordinal numbers or to the corresponding infinite cardinal numbers, as many later mathematicians were to do. Meanwhile, Paul Tannery was stimulated by one of the translations [Cantor, 1883c] to attempt a proof of Cantor's Continuum Hypothesis, which stated that every uncountable set of real numbers has the same cardinality as the set of all real numbers [P. Tannery, 1884]. Unfortunately, this proof was erroneous. The following year he introduced Cantor's work to a philosophical audience (the readers of the *Revue philosophique de la France et de l'etranger*) by describing how Cantor had characterized the notion of a continuum and by characterizing Cantor himself as "a pure idealist as to principles and methods" [P. Tannery, 1885, p. 406]. On the other hand, this Tannery had some reservations about Cantor's infinite ordinal numbers [1885, p. 410]. The following year Jules Tannery published a textbook on the theory of functions, in which he mentioned Cantor's construction of the real numbers, his derived sets, and his discovery that the rational numbers are countable—"a result which appears somewhat paradoxical" [J. Tannery, 1886, pp. x, 43, 58]. Much later he took pleasure in pointing out how his review of 1884 and his textbook of 1886 had "contributed to making G. Cantor's ideas known in France . . ." [J. Tannery, 1901, p. 29].

Several years passed between the contributions of the Tannerys and the next occasion on which a French mathematician utilized Cantor's work. In 1892 Camille Jordan proposed to study the role played by an arbitrary set E in the plane when a real function $f(x,y)$ is integrated over E. With this in mind, he remarked that earlier researchers had presupposed that every such E has a unique measure (*étendue*) and that this measure is finitely additive. That is, if $E_1 \cup E_2 = E$, where E_1 and E_2 are disjoint, then the measure of E equals the sum of the measure of E_1 and the measure of E_2 [1892, p. 69]. After establishing that an arbitrary E has a unique inner Jordan measure and a unique outer Jordan measure, he defined a set in the plane to be Jordan-measurable if its inner and outer Jordan measures are equal. In order to study Jordan measure more deeply, he utilized a number of Cantor's topological concepts such as limit point, derived set, and closed set [Jordan, 1892, p. 72]. When the second edition of Jordan's *Cours d'analyse* appeared the following year, it included both a substantial discussion of these Cantorian concepts and Jordan's theory of measure [1893, pp. 18–30].

Beginning in 1895, a number of French philosophers who were not also mathematicians became interested in Cantor's work. Arthur Hannequin, a young Kantian philosopher, wrote a book criticizing the use of "atoms" in physics and mathematics. Influenced by the philosopher Benno Kerry [1885], Hannequin devoted some twenty-four pages to Cantor and described in some detail the notion of derived set, the infinite ordinals, the characterization of the continuum as perfect and connected, and the status of the Continuum Hypothesis. Although Hannequin was willing to concede the existence of the least infinite ordinal ω, he found Cantor's definition of a continuum to be inadequate, particularly because of its inability to deal with metric considerations such as length: "Thus Cantor's researches have served only to render more obvious the ancient conflict between the continuum on the one hand and the notion of [real] number on the other" [Hannequin 1895, p. 69]. The following year Cantor obtained a more sympathetic hearing from the philosopher Louis Couturat, whose doctoral dissertation [1896] contained seventy-five pages devoted to "the researches, so original and so profound, of Georg Cantor on the theory of infinite sets" [Couturat, 1896, p. 118]. Indeed, Couturat regarded the Continuum Hypothesis as already demonstrated [1896, p. 655], a position that he would modify the following year when reviewing Hannequin's book [Couturat, 1897, p. 781]. It is not surprising, however, that most of the other philosophers who criticized Cantor, such as Georges Lechalas [1897, pp. 627–629] and François Evellin [1898, pp. 115–116], did so by rejecting the actual infinite in favor of the potential infinite.

Until the mid-1890s, the philosophical views of French mathematicians had little effect on their usage of Cantorian notions. Then there emerged three French mathematicians—Emile Borel, René Baire, and Henri Lebesgue—who vigorously applied Cantorian notions while maintaining a philosophical position that would seriously undermine the usefulness of these notions. In this fashion, as we shall see, the contradiction mentioned at the beginning of this section was able to take root.

BOREL, BAIRE, AND LEBESGUE

In 1893 Emile Borel graduated from the *Ecole Normale Supérieure* in Paris. The following year, while teaching at the University of Lille, he completed a doctoral dissertation on the analytic continuation of a complex function, and soon was offered a teaching position at his alma mater. Various measure-theoretic ideas, including the notion of a set of measure zero, were implicit in this thesis [1895]. However, he first developed these ideas in detail within his book *Leçons sur la théorie des fonctions* [1898], which was dedicated to Jules Tannery. The first half of the book was devoted to Cantorian set theory. Reflecting on this period more than a decade later, Borel wrote: "I must admit that, like many young mathematicians, I was at first seduced by the Cantorian theory. I do not regret this, for it is a subject which renders the mind singularly flexible" [1912, p. 2].

Borel's monograph developed those parts of set theory which he deemed useful in complex analysis, in particular, elementary cardinal arithmetic and certain topological notions. Although he expressed some reservations about cardinal numbers greater than c (the power of the set of all real numbers), he accepted various theorems of Cantor's, such as the following:

> If a family A of sets has power c and if every set in A has power c, then the union of A also has power c [Borel, 1898, pp. 16–17]. **(1)**

If Borel had carefully examined his proof of this theorem, he would have found that it depended on the arbitrary choice, for each set B in A, of a one–one function from B onto c. The use of such arbitrary choices, which was becoming widespread late in the nineteenth century, raised few doubts at the time.[c] Yet such choices concealed a potent new mathematical assumption that was not formulated explicitly until 1904: the Axiom of Choice. This axiom states that if A is a family of non-empty sets, then there exists a function f such that $f(B) \in B$ for every set B in A. In more psychological terms, such a function f "chooses" an element from each set B in A. What remained unclear was how to specify such a function uniquely for a given family A.

It was this fact which made Borel an opponent of the Axiom of Choice when it was first stated explicitly in 1904. Indeed, Borel held definite philosophical views on the subject of what kind of mathematics was legitimate. In his opinion, mathematical objects had to be determined by a *rule,* that is, by a construction that uniquely specifies the object constructed. Yet in his monograph, unbeknownst to himself, Borel had repeatedly violated his own philosophical dictum by implicitly using the Axiom of Choice at several junctures, including theorem (1) above.

Eventually, although this was not apparent at first, the Axiom of Choice would greatly affect the theory of measure that Borel developed in his monograph. This theory of measure, he stressed, originated from quite different problems than had Jordan's theory. Restricting himself to the real interval [0,1], Borel wrote:

> When a set consists of all the points in a denumerable infinity of disjoint intervals having total length s, we say that the set *has measure s.* When two disjoint sets have measures s and s', their union has measure $s + s'$. . . . More generally, if we have a denumerable infinity of disjoint sets having respectively the measures $s_1, s_2, \ldots, s_n, \ldots$, their union . . . has measure
>
> $$s_1 + s_2 + \cdots + s_n + \ldots$$
>
> [countable additivity]. If a set E has measure s and contains every point in a set E' whose measure is s', *the set $E - E'$ . . . will be said to have measure $s - s'$ . . . . The sets that can be defined by the preceding definitions will be said to be measurable sets . . .* [1898, pp. 47–48].

In this fashion Borel introduced a class of measurable sets, which, following later usage, we shall call the Borel sets. At the time he did not develop their theory much further, except to prove that every closed set is a Borel set and that every countable set, as well as some uncountable sets, has measure zero.[d] It could easily be shown, he insisted in a note at the end of his book, that the

[c]At the time such choices raised doubts only in the mind of Giuseppe Peano [1890, p. 210] at Turin, and subsequently in the minds of two of his colleagues, Rodolfo Bettazzi [1896] and Beppo Levi [1902]. For a discussion of this matter, see Moore, 1982, pp. 26–29 and 76–82.

[d]In the present context a set is said to be finite if it is empty or has the same cardinality as the set $\{1, 2, \ldots, n\}$ for some natural number n; to be denumerable if it has the same cardinality as the set of all natural numbers; to be countable if it is either finite or denumerable; and to be uncountable if it is not countable.

following proposition is true:

> The set $\mathcal{B}$ of all Borel sets has the power c of the
> continuum [1898, p. 110]. (2)

Much later, however, mathematicians discovered that if the Axiom of Choice were false, it could happen that the set $\mathcal{B}$ has a power greater than c, that *every* set of real numbers is a Borel set, and that the hierarchy of the Borel sets collapses to the first three levels beyond the closed sets.[e]

The admiring but somewhat ambivalent attitude that Borel held toward Cantor can be seen in two articles which Borel wrote for the *Revue philosophique*. The first of these, published in October 1899, was stimulated by François Evellin's article [1898] in the same journal. Evellin had objected not only to Cantor's sequence of infinite cardinals of ever higher power but even to his use of the actual infinite at all. He wrote: "If, in order to define the irrational numbers, one introduces the notion of *set* in Cantor's sense, then one acts in a futile and dangerous manner ..." [Evellin, 1898, p. 116]. In fact, Evellin asserted, the notion of the actual infinite is betrayed by an internal contradiction. Borel, on the other hand, approved of Cantor's notion of denumerable set (and hence of the actual infinite) but held certain reservations about the transfinite (*i.e.*, uncountable cardinals and ordinals):

> When, some twenty years ago, G. Cantor published his ideas on the enumeration of numbers larger than the infinite, the mathematical public greeted them mistrustfully. Nevertheless, many analysts were struck by the beauty of these speculations and did not let themselves be stopped by the form, intentionally paradoxical in places, of Cantor's exposition. ... His new ideas have now ... become classic, both for the mathematicians and for the philosophers [Borel, 1899, p. 383].

The remainder of this article was devoted to discussing du Bois-Reymond's theorem that, given any sequence of increasing real functions, there is a real function that increases still faster. By way of conclusion, Borel offered some reflections on the significance of that theorem for Cantor's work:

> One may ... ask, quite simply, whether we can have a conception of the uncountable sequence of ever larger increasing functions. The difficulty in acquiring this conception comes from the fact that any *determinate* process of forming this sequence, since it is expressed by means of a finite number of words, leads only to a denumerable sequence. But, on the other hand, the theorem of Paul du Bois-Reymond tells us how to exceed this denumerable sequence once it is formed. Here is an antinomy which analysis appears incapable of resolving. It seems clear, however, that the subsequent development of mathematics will necessarily lead to a choice—to accept or to proscribe the *transfinite* (or uncountable infinity) [Borel, 1899, pp. 389–390].

In a second philosophical article, published the following year, Borel

---

[e]See Feferman and Levy, 1963, and the discussion in Moore, 1982, pp. 102–103. In particular, there is a model of Zermelo-Fraenkel set theory in which every set of real numbers is a countable union of countable sets.

expressed his reservations about the uncountable even more directly: "It seems to me that those who refuse to consider the notion of transfinite to be *as clear at present* as that of the indefinite [*i.e.,* the denumerable] are in the right" [1900, p. 382]. With the appearance in 1904 of Zermelo's Axiom of Choice, Borel's reservations would become much more pronounced.

Like his compatriot Borel, René Baire attended the *Ecole Normale Supérieure.* After graduating in 1895, Baire taught in various *lycées* and simultaneously completed his doctoral thesis, which was published in 1899. As Lebesgue later pointed out [1922, pp. 16–17], Baire "was the first to devote all his mathematical activity to the theory of real functions. . . ." Until the work of Baire, nineteenth-century analysts—particularly those in France— concerned themselves principally with the theory of complex functions. Baire's researches, which were influenced by the Italian analyst Vito Volterra, influenced Lebesgue in turn, and thereby brought about the flowering of the modern theory of real functions.

Baire shared with Borel a somewhat ambivalent attitude toward Cantorian set theory. In the conclusion to his thesis, which relied heavily on Cantor's results, Baire wrote:

> The theory of point-sets plays a very important role. . . . One can even say . . . that, from the perspective which we have adopted, every problem in the theory of [real] functions leads to certain problems in set theory; and it is to the extent that progress is made, or can be made, on these latter questions that it is possible to resolve more or less completely the given problem [1899, p. 121].

At the same time Baire refused to grant the existence of transfinite ordinals, which, however, he continued to use as superscripts of derived sets:

> On this subject I remark once and for all that we shall not be concerned with the difficulties that the abstract notion of *transfinite* [*ordinal*] *number* may involve, even though we may employ this expression in the rest of this work. In the present case, for example, the set P^α (where α is a definite number in the second number-class) represents something perfectly definite, independent of any abstract considerations relative to Cantor's symbols. Thus, in the usage that we make of the term *transfinite number* there is nothing more than a convenient language [1899, p. 36].

In his dissertation Baire introduced a hierarchy of classes of real functions. The essential feature of his hierarchy was that it treated discontinuous functions as the iterated limit of sequences of continuous functions. More precisely, he defined Baire class zero to be the set of all continuous real functions; for any other countable ordinal α, Baire class α was defined to be the set of all those real functions f, not in any previous class, such that

$$f(x) = \lim_{n \to \infty} f_n(x)$$

for every real x, where $f_1, f_2, \ldots$ was some sequence of real functions in previous classes. Thus, in particular, Baire class one was the set of all real functions that are the pointwise limit of a sequence of continuous functions. As a result, Cantor's ordinal numbers were essential to Baire's approach.

Moreover, by way of a framework for his classification, he explicitly accepted Dirichlet's general definition of a real function as any correspondence $x \rightarrow f(x)$ between real numbers. "In this definition," Baire observed, "it is beside the point to ask how the correspondence can be determined effectively or whether it is even possible to determine it effectively" [1899, p. 1]. This easy tolerance of non-constructive methods disappeared from Baire's work in the aftermath of Zermelo's introduction of the Axiom of Choice.

In order to derive results about his classification of real functions, Baire indirectly relied on the Axiom of Choice several times. One of the propositions that served as a conduit for such indirect uses of this axiom was the Countable Union Theorem, due to Cantor: The union of a countable family of countable sets is countable [1878, p. 243; 1883c, p. 313]. This theorem proved useful in establishing the basic properties of sets of first and second category, introduced by Baire the year before he published his thesis. Baire defined a set A of real numbers to be of first category if A is the union of countably many nowhere dense sets; otherwise, A was said to be of second category [1898]. Baire's thesis relied on the Countable Union Theorem to obtain the following elementary result:

> The union of countably many sets, each of first
> category, is also of first category [1899, p. 65].   (3)

Furthermore, he used the Countable Union Theorem to establish that no new functions would be added to his classification by extending it beyond the countable ordinals [1899, p. 70]. Finally, he relied implicitly on the Axiom to show that this classification did not exhaust all the real functions, since the set of functions in classification had cardinality c while the set of all real functions had a higher cardinality [1898, p. 1623]. If the Axiom of Choice were false, then all of Baire's results mentioned here could be false as well.

It is intriguing that Baire, who largely shared Borel's constructivist philosophy of mathematics, had a premonition that non-constructive methods might affect his own research. At first, Baire centered his investigations on Baire classes one and two. On 25 October 1898 he wrote to Volterra, whom he had visited in Italy not long before, in order to express his uncertainty about the existence of Baire class three and higher classes. Furthermore, Baire wondered whether it was possible to define a real function escaping from his classification:

> We would know how to obtain a function *beyond* all the [Baire] classes if we knew how to partition the continuum [of real numbers] into two sets, each of second category on every interval. We would let $f = 0$ for the first set and $f = 1$ for the second. I have not succeeded in *defining* such a partition, and I have come to wonder whether it is even *possible* to *define* such a partition. Clearly, it is a delicate matter not only to answer but even to pose such a question. The meaning of the phrase "to define a set" would have to be made more precise. Yet I am convinced that the act of imposing on a set [of real numbers] the condition of being definable must restrict the notion of set to a considerable degree. ... I would not be surprised if it were impossible, with the means that we have

available, to define a property partitioning the real numbers into two sets A and B such that both sets are of second category on every interval. We would have a proposition like the following, which is rather unexpected: If a set A and its complement B can be defined, then there exists a closed *nowhere dense* set P such that either A or B is of first category on any interval not intersecting P. . . . All of this is closely related to Borel's reflections in the notes at the end of his book on functions [1898]. . . . May we not hope to learn in this way how far we are permitted to use the notion of *arbitrary real function?* [Baire, quoted in Dugac, 1976, pp. 347–348]

In a sense, Baire's qualms were justified. The existence of such a partition as described in this quotation would imply the existence of a set of real numbers lacking what was later called the Baire property. By definition, a set A of real numbers has the Baire property if there is no interval such that A and its complement both intersect the interval in sets that are of second category. Around 1900, Baire pointed out to Lebesgue that it would be very interesting to define a set of real numbers that lacked the Baire property and hence to define a real function escaping from Baire's classification [Lebesgue, 1905, p. 186]. As later became clear, however, the existence of a set of real numbers lacking the Baire property, of a real function not in Baire's classification, and even of a real function not belonging to Baire class zero, one, or two, depend on the Axiom of Choice.

As Borel and Baire had done earlier, Henri Lebesgue attended the *Ecole Normale Supérieure*. After graduating in 1897, Lebesgue published a number of short papers, one of which generalized some of Baire's results to functions of several real variables [Lebesgue, 1899]. Shortly afterward, he made his first contribution to measure theory and gave the first version of his Measure Problem:

The Measure Problem for plane regions bounded by simple closed curves can be posed in the following manner . . . : To associate with each such region a number, to be called *area,* such that congruent regions have equal areas and such that the region formed from the union of finitely or infinitely many regions, which have part of their boundary in common and which do not overlap, has as its area the sum of the areas of the component regions . . . [1899a, p. 870].

Here, in effect, Lebesgue adopted Borel's proposal that a measure should be countably additive.

Lebesgue's doctoral dissertation [1902] elaborated his new theory of measure and integration.[f] There he presented his notion of measurable set, which generalized the Borel sets: If A is a bounded set of real numbers in some fixed interval I of length k, the outer measure of A is defined to be the least upper bound of the sum of the lengths of countably many intervals covering A; the inner measure of A is defined to be $k - b$, where b is the outer measure of $I - A$. If the outer measure of A and the inner measure of A are equal, then A is said to be measurable [1902, pp. 237–238].

It was specifically in order to solve what he called the Measure Problem

---

[f]For an incisive historical treatment of Lebesgue's theory and its origins, see Hawkins, 1975.

that Lebesgue introduced his measurable sets. He posed this problem in a form that was quite general for his time:

> We propose to assign to each set [in n-dimensional space] a non-negative [real] number, which we shall call its measure, satisfying the following conditions:
>
> (i) There is a set whose measure is not zero.
>
> (ii) Congruent sets have equal measure.
>
> (iii) The measure of the sum [union] of a finite or denumerable infinity of disjoint sets is the sum of the measures of these sets.
>
> We shall resolve this *Measure Problem* only for the sets that we call measurable.[g]

To show that the measurable sets of real numbers are countably additive (condition (iii)), he first had to establish that the union of denumerably many disjoint measurable sets is measurable. He began by supposing that $A_1, A_2, \ldots$ is a sequence of disjoint measurable sets included in some real interval [a, b]. Next, he enclosed each set A_i in the union of some countable family $\mathcal{F}_i$ of open intervals, and each set [a, b] $- A_i$ in the union of another countable family $\mathcal{G}_i$ of open intervals, in such a way that each set $E_i = (\cup \mathcal{F}_i) \cap (\cup \mathcal{G}_i)$ had a total length d_i; here the families were chosen so that the sum of the d_i was an arbitrarily small positive number [1902, p. 239]. Nevertheless, Lebesgue provided no rule for choosing the families $\mathcal{F}_i$ and $\mathcal{G}_i$ for each i. Nor could he have done so, thereby avoiding arbitrary choices, since the Axiom of Choice is necessary to ensure the countable additivity of Lebesgue measure. More than a decade later, thanks to the researches of Wacław Sierpiński [1916; 1918], Lebesgue was faced with the dilemma of either accepting a limited form of the Axiom of Choice or else restricting his theory of measure and integration in an essential way. For, if the Axiom of Choice were false, it could happen that Lebesgue measure is not countably additive.[h]

From the additivity of his measure, Lebesgue concluded that every Borel set is measurable. Since, Lebesgue argued, there are only as many Borel sets as there are real numbers, and since there is a perfect nowhere dense set of measure zero, say A, then there is a measurable set that is not a Borel set. For the subsets of A are all measurable, being of measure zero, and the set of these subsets has a power greater than that of the real numbers, since A is perfect [Lebesgue, 1902, pp. 240–241]. Nevertheless, if the Axiom of Choice were false, then there could be just as many Borel sets as measurable sets, but there could also be a Borel set that is non-measurable.

Lebesgue also asserted that every real function in Baire's classification is a measurable function, where a real function f is said to be measurable if $\{x: a < f(x) < b\}$ is a measurable set for every real a and b. Indeed, Lebesgue showed that the sum of two measurable functions is measurable, and similarly for the product. However, when he proved that the pointwise limit f of a sequence f_i of measurable functions is measurable (provided that f is

---

[g]Lebesgue, 1902, p. 236. In his book on integration [1904, p. 103], he replaced condition (i) by the equivalent requirement that the n-dimensional unit cube have measure one.

[h]This is the case in the Feferman-Levy model mentioned in footnote e.

bounded), he used the proposition that the intersection of a countable family of measurable sets is measurable, and so relied indirectly on the countable additivity of Lebesgue measure. Hence his proof, whose immediate consequence was that every function in Baire's classification is measurable [Lebesgue, 1902, p. 257], used the Axiom of Choice in an essential way.

The measurable sets might not solve the Measure Problem, Lebesgue observed, since there might exist a bounded set of real numbers that was non-measurable. Moreover, he entertained the possibility that the Measure Problem might have a solution even if there existed a non-measurable set: "It has not been shown in any way that the Measure Problem is unsolvable for those sets (if any exist) whose outer and inner measures are unequal" [1902, p. 239]. In his book on the theory of integration Lebesgue maintained this viewpoint [1904, p. 106]. Furthermore, in an article published in 1905 but written early the previous year, he attempted to define a specific measurable set that is not a Borel set (not realizing that arbitrary choices had crept into his proof), and concluded by noting: "But the much more interesting question— can one specify a non-measurable set?—remains unanswered" [1905, p. 216]. The existence of such a non-measurable set turned out to be intimately related to Zermelo's Axiom of Choice.

ERNST ZERMELO AND THE WELL-ORDERING PROBLEM

In 1883 Cantor introduced the notion of a well-ordered set, that is, a linearly ordered set such that every non-empty subset has a least element.[i] Impressed by the properties of well-ordered sets and especially by the alephs (the cardinal numbers of infinite well-ordered sets), he asserted as a "law of thought" the proposition that every set can be well-ordered. By 1895 he had come to believe that this proposition required a proof. Not long afterward, he found an argument for this proposition and sent it to David Hilbert in 1896 or 1897, then to Richard Dedekind in 1899. However, Cantor would not permit this argument to be published, even though Philip Jourdain, who had found a similar argument [1904], asked permission to do so [Grattan-Guinness 1971, p. 115].

In August 1904 Julius König, a Hungarian mathematician, gave a lecture to the International Congress of Mathematicians at Heidelberg and purported to show that the set of all real numbers cannot be well-ordered. Cantor, who was present at König's lecture, could not find the gap in the argument. By the following day, however, Zermelo had isolated this gap [Kowalewski, 1950, p. 202].

Zermelo set to work in order to establish that Cantor's claim was correct: every set can be well-ordered. A month after König's lecture, Zermelo succeeded [1904]. He based his demonstration of this Well-Ordering Theo-

---

[i]Cantor, 1883a, p. 550. Strictly speaking, Cantor did not state this definition at the time, but one that was equivalent to it and considerably more complicated.

rem on a new postulate which soon came to be known as Zermelo's Axiom, or the Axiom of Choice. The proof gave rise to a controversy that, over the next four years, involved mathematicians in England, France, Germany, Holland, Hungary, Italy, and the United States [see Moore, 1978 and 1982].

This controversy entered France primarily via Hilbert, then an editor of *Mathematische Annalen,* to whom Zermelo had first sent his proof. On behalf of the *Annalen,* Hilbert asked Borel to prepare a reply to Zermelo's proof. Completed on 1 December 1904, Borel's brief article rejected the proof. What Zermelo had actually shown, Borel noted, was that the following two problems are equivalent for a given set M:

(A) to well-order M;
(B) to choose an element from every non-empty subset of M.

What Zermelo had *not* done, Borel insisted, was to prove that the equivalence of (A) and (B) provides a general solution for problem (A). To resolve (B) required "a means, at least a theoretical one, for determining a distinguished element m' from an arbitrary subset M'" of M [1905, p. 194]. In particular, Zermelo had supplied no such means for arbitrary sets of real numbers and hence had not shown that the set of all real numbers can be well-ordered. Borel closed his article by quoting part of a letter that he had received from Baire on the same subject. Baire doubted that it would ever be possible to well-order the set of all real numbers, since he considered this set and Cantor's infinite ordinals to be mere potentialities.

Borel's article stimulated a lively exchange of letters among those French mathematicians who at the time were applying Cantorian notions to other parts of mathematics. Writing to Borel, the analyst Jacques Hadamard disagreed with the conclusions of Borel's article. Hadamard accepted Zermelo's proof of the Well-Ordering Theorem because the proof used simultaneous choices, rather than successive choices depending on those made previously. Borel had claimed that "any argument which supposes an *arbitrary choice* to be made uncountably many times ... [is] outside the domain of mathematics" [1905, p. 195]. Hadamard, on the other hand, found uncountably many arbitrary choices to be quite as acceptable as countably many. For Hadamard, the essential distinction was not one of cardinality, but the difference between showing that there exists a function with a given property and that a function with this property can be specified uniquely [Baire *et alii,* 1905, pp. 261–263].

Borel forwarded Hadamard's letter to Baire, who elaborated on his own previous letter. On the whole, Baire agreed with Borel, but considered himself to be more radical when he regarded every infinite set as merely a potential infinity established by convention. For a given infinite set A, Baire insisted, it was false to regard every subset of A as given. Consequently it made no sense to assume, as Zermelo had done, that an element was chosen from every non-empty subset of A. Zermelo's Axiom, Baire continued, was not contradictory but simply meaningless; and, in the final analysis, everything in mathematics must be reduced to the finite [Baire *et alii,* 1905, pp. 263–264].

Borel next solicited Lebesgue's opinion. Although at first he seemed to take the middle ground between Borel and Hadamard, Lebesgue gradually formulated a distinct but subtle position rejecting the Axiom of Choice. For Lebesgue, the principal question was whether one can prove the existence of a mathematical object without defining it. He admitted that it was a matter of convention how one answered this question, and conceded that sometimes he had published demonstrations in which he did not define a unique object of the type shown to exist. Nevertheless, he remained convinced that one could prove the existence of a given kind of mathematical object *only* by defining a unique particular instance of such an object. It was not enough to show that the class of all such objects was non-empty. In conclusion, Lebesgue differed from Borel by rejecting the Axiom of Choice not only for uncountable families of sets but for denumerable families as well [Baire *et alii,* 1905, pp. 264–269].

A NEGATIVE SOLUTION TO LEBESGUE'S MEASURE PROBLEM

In 1905, not long after this exchange of letters was published, an Italian mathematician applied the Axiom of Choice to Lebesgue's Measure Problem. Giuseppe Vitali, who had discovered the measurable sets independently of Lebesgue, familiarized himself with Lebesgue's book of 1904 on integration. By using the Axiom of Choice, Vitali established that there exists a subset of the real line which is not Lebesgue-measurable, *i.e.,* a non-measurable set. Vitali had a broader goal, however: to prove that Lebesgue's Measure Problem has no solution. In particular, he set out to show that there is no translation-invariant, countably additive, positive real measure defined on all bounded subsets of the real line and such that the unit interval has measure one.

Vitali began his proof by supposing that there was such a measure. For each real x, he let

$$A_x = \{x + b: b \text{ is rational}\}$$

and then designated H as the set of all such A_x. Selecting an element p in $A_x \cap (0, \frac{1}{2})$ for each distinct A_x in H, an act justified by the Axiom of Choice, he termed G_0 the set of all such p. Then he defined

$$G_q = \{p + q: p \in G_0\}$$

for each rational number q. Since the G_q are disjoint congruent sets, they all have the same measure. Moreover, each of $G_0, G_{1/2}, G_{1/3}, \ldots$ is a subset of the closed interval $[0, 1]$ and hence the sum of their measures is at most one. Yet by the countable additivity of the measure, each of them has measure zero. As a result, the measure of G_q is zero for every rational q, and so the measure of the real line (the union of all the G_q) is likewise zero—a contradiction. Consequently, Lebesgue's Measure Problem had no solution. Lebesgue's particular measure did not apply to every bounded set of real numbers, and so did not fulfill all the conditions that he had demanded of it. Vitali concluded by noting that the set G_0 required some comment. This set certainly existed if the real numbers could be well-ordered. Those who did not grant the existence of such a well-ordering, he added, could conclude instead that there cannot be both a solution to the Measure Problem and a well-ordering of the real line [1905, pp. 3–5].

Two years later Lebesgue himself supplied an example of a set that is not Lebesgue-measurable, and simultaneously revealed how his philosophical presuppositions affected his attitude toward the existence of such a set. During 1905 he had written a monograph (not published until 1971) in which he adopted a distinction that had first been introduced some years earlier by Paul du Bois-Reymond [1887, p. 64], the distinction between "Idealists" and "Empiricists." According to Lebesgue, the Empiricists admitted the existence only of those real functions that could be uniquely defined, whereas the Idealists accepted the existence of real functions whose existence had been "proved" in some sense but which could not be uniquely defined [Lebesgue, 1971, pp. 37–39]. It is clear from the context that he regarded Zermelo and Hadamard as Idealists, while he considered Baire, Borel, and himself to be Empiricists. When he provided an example of a non-measurable set in 1907, and credited Vitali with an earlier example, Lebesgue insisted that its existence was established for Idealists, who accepted the Axiom of Choice, but not for Empiricists, who rejected this axiom [1907a, pp. 32–39]. Indeed, Lebesgue continued to doubt that a non-measurable set of real numbers would ever be defined uniquely [1907, pp. 202–203].

In 1914 Felix Hausdorff, a German-Jewish astronomer turned set theorist, investigated Lebesgue's Measure Problem in greater depth. Hausdorff, who accepted the Axiom of Choice completely, published an article [1914] in which he modified that problem. As a first step, he gave a demonstration somewhat different from Vitali's that the Measure Problem has no solution for the real line and hence lacks a solution for n-dimensional space whenever $n \geq 1$. He then inquired whether this problem would have a solution if the requirement (iii) of countable additivity were weakened to that of finite additivity:

(iv) $m(A \cup B) = m(A) + m(B)$ if A and B are disjoint.

While he was unable to resolve this weakened problem for $n = 1$ or $n = 2$, he succeeded in proving that there is no solution whenever $n \geq 3$.

Hausdorff's proof, which extended earlier arguments yielding a set that is not Lebesgue-measurable, showed that the existence of a solution for $n = 3$ would imply that one half of a sphere is congruent to one third of the same sphere. His strategy was to assume that there existed a measure satisfying (i), (ii), and (iv), and then to decompose a given sphere S into four sets A, B, C, D such that A, B, C, and $B \cup C$ were congruent to each other while D had measure zero. To obtain this decomposition, he let ϕ and ψ be rotations of S, through 180 and 120 degrees respectively, on distinct axes passing through the center of S. With ϕ, ψ, and ψ^2 treated as different formal factors, a group G of motions was generated. Each member of G had a unique representation either as ϕ or as

$$\phi^{i_1}\psi^{j_1}\phi^{i_2}\psi^{j_2}\ldots\phi^{i_{n-1}}\psi^{j_{n-1}}\phi^{i_n}$$

for some n, where i_1 and i_n equaled zero or one, i_k equaled one if $1 < k < n$, and j_k equaled one or two. (What Hausdorff had obtained here, although he did not speak in these terms, was a free group with two generators.) The set D of those points fixed under members of G was denumerable, and was shown to have measure zero.

Finally, Hausdorff determined the sets A, B, and C. For each point x of $S - D$, he defined $P_x = \{g(x) : g \in G\}$, and thus for every x and y in $S - D$, P_x and P_y were equal or disjoint. Using the Axiom of Choice to select a point from each distinct P_x, he called the resulting choice set M. Then $S - D$ was the union of the images of M under all g in G, and he partitioned $S - D$ into three sets A, B, and C by means of these images: For each g in G, exactly one of the two images $g''M$ and $(g\phi)''M$ was included in A and the other in $B \cup C$, while exactly one of the three images $g''M$, $(g\psi)''M$, and $(g\psi^2)''M$ was included in A, one in B, and one in C. It followed that A, B, C, and $B \cup C$ were all congruent to each other since $\psi''A = B$, $(\psi^2)''A = C$, and $\phi''A = B \cup C$. But then A had the measure of one half and of one third of the same sphere, a contradiction, completing the proof [1914, pp. 430–433].

Thanks to Borel, this result soon became known as Hausdorff's paradox. In the second edition [1914] of his book on set theory and complex functions, Borel included an exposition of Hausdorff's counterintuitive result, and concluded that the culprit was the Axiom of Choice:

> If, then, we designate by a, b, c, the probability that a point in S belongs to A, B, or C respectively and if we grant that the probability of a point belonging to a set E is not changed by a rotation about a diameter (this is what Lebesgue expresses by saying that two congruent sets have the same measure), one obtains the contradictory equalities: $a + b + c = 1$, $a = b$, $a = c$, $a = b + c$.
> The contradiction has its origin in the application . . . of Zermelo's *Axiom of Choice*. The set A is homogeneous on the sphere; but it is at the same time a half and a third of it. . . . The paradox results from the fact that A *is not defined*, in the logical and precise sense of the word *defined*. If one scorns precision and logic, one arrives at contradictions [Borel, 1914, pp. 255–256].

At last, Borel was convinced, his decade of opposition to the Axiom of Choice had been fully vindicated. Since this axiom generated a contradiction, its illogical character could not help but be obvious to mathematicians. In time, Borel's view came to be shared by a number of mathematicians who regarded Hausdorff's paradox (or its successor, the Banach-Tarski paradox) as sufficient reason to be wary of the Axiom of Choice.

LEBESGUE AND SIERPIŃSKI

It was Wacław Sierpiński who first realized how deeply the Axiom of Choice was entangled with Baire's classification of functions and with Lebesgue's theory of measure. On 4 December 1916, Emile Picard submitted to the Paris

Academy of Sciences a brief note by Sierpiński on the role of the Axiom of Choice in real analysis. There Sierpiński pointed out how the Denumerable Axiom of Choice (*i.e.,* the Axiom of Choice restricted to denumerable families of sets) was used in Baire's proof [1899, p. 69] that every function f in Baire class two is the iterated limit of some double sequence $f_{n,k}$ of continuous functions. Furthermore, Sierpiński observed that Lebesgue had implicitly used the Denumerable Axiom to show that Lebesgue measure is countably additive [Sierpiński, 1916, p. 690].

The following year the Paris Academy of Sciences printed another note by Sierpiński, who demonstrated that certain seemingly innocuous assumptions (having no obvious connection with the Axiom of Choice) implied the existence of a non-measurable set and hence depended on that axiom. In particular, he established that there is such a non-measurable set if the set of all real functions can be linearly ordered, or if the family of all denumerable sets of real numbers has the power c, or even if Baire class two has power c [1917].

Lebesgue [1918, p. 238] responded to Sierpiński's second article by claiming that Sierpiński had inadvertently used the Axiom of Choice to show that the existence of a non-measurable set follows from the proposition that the family of all denumerable sets of real numbers has power c. Nevertheless, Lebesgue was mistaken. As for Sierpiński's first article and, in particular, his assertion that the countable additivity of Lebesgue measure depends on the Denumerable Axiom, Lebesgue did not reply. In 1934 Charles de la Vallée-Poussin, another French analyst whose constructivistic views paralleled Lebesgue's, made a valiant but ultimately futile attempt to prove the countable additivity of Lebesgue measure without using the Axiom of Choice [de la Vallée-Poussin, 1934, p. 26].

In 1918 Sierpiński published a lengthy article analyzing the relationship between the Axiom of Choice and many propositions used in real analysis or set theory. Thereby he became the first to explain just how deeply the researches of Baire, Borel, and Lebesgue depended on that axiom. In order to understand Sierpiński's results, it will be helpful to return briefly to the article that Lebesgue wrote during 1904 and published the following year. In that article Lebesgue investigated a class of real functions which he called analytically representable. A real function f was said to be analytically representable if f is obtained from countably many variables and constants by the operations of addition, multiplication, and taking the limit of a sequence of functions. He argued that a real function f is in Baire's classification if and only if f is analytically representable [1905, p. 152]. After defining f to be a Borel function if and only if the set $\{x: a \le f(x) \le b\}$ is a Borel set for every real number a and b, he demonstrated that f is analytically representable if and only if f is a Borel function [1905, p. 168]. Furthermore, from the countable additivity of Lebesgue measure he deduced that every Borel set is measurable and that every function in Baire's classification is a measurable function. Finally, Lebesgue supplied an example of a real function that is measurable but not in Baire's classification, and of a measurable set that is not a Borel set [1905, pp. 212–216].

Sierpiński [1918] recognized that Lebesgue's results were thoroughly entangled with the Axiom of Choice. In particular, Sierpiński observed, this axiom was used indirectly to ensure that every function in Baire's classification is analytically representable and hence measurable. On the other hand, he remarked that the Axiom of Choice appeared to be necessary to obtain the existence of a real function not in any Baire class or even of a function beyond Baire class three. In the absence of this axiom, he added, it was not known how to show that if $f_1(x), f_2(x), \ldots$ is a convergent sequence of functions in Baire's classification, then the function $f(x) = \lim_{n \to \infty} f_n(x)$ also belongs to Baire's classification, for such a demonstration relied on Cantor's Countable Union Theorem that the union of a countable family of countable sets is countable, a theorem that in turn depended on the Axiom of Choice [Sierpiński, 1918, pp. 133–137].

As a particularly important case of the Countable Union Theorem, Sierpiński focused on the following proposition:

> The set of all real numbers is not a countable union
> of countable sets.   (4)

In fact, he continued, the countable additivity of Lebesgue measure implied (4), and so did the apparently weaker proposition that the union of countably many sets having Lebesgue measure zero also has Lebesgue measure zero. Moreover, the Countable Union Theorem yielded a particular case of the Axiom of Choice, namely, the Denumerable Axiom restricted to families containing only countable sets [1918, pp. 113–114].

If (4) were false, then the hierarchy of Borel sets would be trivialized, as would Baire's classification of real functions. Instead of extending through all the countable ordinals (and even then including only a negligible portion of all sets of real numbers), the Borel hierarchy would be reduced to the first four levels. Every set of real numbers would be Δ_4^0 in modern notation, that is, a countable union of a countable intersection of a countable union of closed sets, as well as the complement of such a set. Every real function would be in some Baire class. Furthermore, every measurable set would be a Borel set, but Lebesgue measure would not be countably additive.

Not until 1963 was it shown that there is a model of Zermelo-Fraenkel set theory in which (4) is false. Using Paul Cohen's [1963] method of forcing, Solomon Feferman and Azriel Levy [1963] obtained such a model in which the Axiom of Choice is strongly violated. Thanks to recent research by Juris Steprāns, it is now known that in the Feferman-Levy model there is a Borel set that is non-measurable. In this way Sierpiński's intuition about the role of the Axiom of Choice in the researches of Baire, Borel, and Lebesgue was fully corroborated.

As late as 1938, Lebesgue actively continued to oppose the Axiom of Choice. In December of that year, a conference on the foundations of mathematics was held in Zurich. Among the participants were Lebesgue and Sierpiński, who gave consecutive lectures about the Axiom of Choice. Lebesgue congratulated Sierpiński and his students on their extensive researches on the axiom. All the same, Lebesgue remained a subtle adversary of this assumption. When the controversy over the axiom began in 1905, he observed, the opposing sides could not make themselves understood, because they lacked a common logic; indeed, logic itself was at stake [Lebesgue, 1941, p. 116].

Yet Lebesgue knew quite well that the issues involved were not limited to the underlying logic. Despite the fact that during three decades of active use the Axiom of Choice had not led to a contradiction, there remained a sense of unease. For logic did not create confidence in a proposition, he insisted, unless one had already accepted the proposition as plausible. By way of conclusion, he argued that "in the studies on the foundations and methods of mathematics, there must be a large place for psychology and even for esthetics" [1941, p. 122]. Thereby he underlined the psychologistic perspective that he had shared with Baire and Borel at least since 1905.

What is surprising in this controversy is how little effort Baire, Borel, and Lebesgue appear to have made to understand precisely how the Axiom of Choice affected their discoveries. Lebesgue, in particular, seems never to have attempted seriously to grasp how essential this axiom was to the countable additivity of his measurable sets and hence to the countable additivity of the Lebesgue integral. Only Leonida Tonelli in Italy endeavored to develop a version of the Lebesgue integral without the Axiom of Choice and consequently without countable additivity. However, Tonelli's research [1921] attracted little interest.

GENERALIZING LEBESGUE'S MEASURE PROBLEM

Hausdorff's research of 1914 on the Measure Problem was continued by a young Polish mathematician, Stefan Banach, who had recently written his doctoral thesis [1922] at Lwow on linear operators and who was soon to develop a school of functional analysis there. Banach published in 1923 an article devoted to what he called Hausdorff's Extended Measure Problem:

> To find a non-negative function m, defined on every bounded subset A of n-dimensional Euclidean space, such that
> (1) $m(A)$ is positive for some A.
> (2) Congruent sets have equal measures.
> (3) $m(A \cup B) = m(A) + m(B)$ if A and B are disjoint.

Like the majority of Polish mathematicians, Banach accepted the Axiom of Choice. Consequently, he accepted as well Hausdorff's result that the Extended Measure Problem has no solution for $n \geq 3$.

All the same, the Extended Measure Problem remained open for the cases $n = 1$ and $n = 2$: the Euclidean line and plane. By means of the Well-Ordering Theorem Banach established that the Extended Measure Problem has a positive solution in these two cases [1923, p. 19]. Thus an additive, congruence-preserving measure could be defined on all subsets of the line, and another such measure on all subsets of the plane. Moreover, in each case this measure could be assigned so as to agree with Lebesgue measure for all

Lebesgue-measurable sets, or so as to assign the measure one to some set with Lebesgue measure zero. As a result, Banach obtained two types of integral, each of which was defined on all subsets of the line or of the plane. One of these extended the Lebesgue integral, whereas the other extended the Riemann integral but differed at times from the Lebesgue integral. Both of Banach's integrals were additive but, necessarily, neither was countably additive.[j]

Because of the disparity of the Extended Measure Problem between the cases $n = 1$ and $n = 2$ on the one hand and $n \geq 3$ on the other, Banach attempted to come to a deeper understanding of Hausdorff's proof for the latter case. During the same period Alfred Tarski, one of Sierpiński's former students at Warsaw, undertook some related research. Banach and Tarski wrote a joint paper showing, among other things, that it is impossible to partition a plane polygon P into a finite number of pieces and reassemble them to obtain a polygon strictly including P [1924]. By contrast, they established that in three dimensions a polyhedron—and indeed *every* bounded set with non-empty interior—can be partitioned into a finite number of pieces and reassembled to give any other bounded set with non-empty interior. Thus, in particular, a sphere of unit radius could be partitioned into a finite number of pieces and reassembled into two spheres, each of unit radius. This implausible result became known to later mathematicians as the Banach-Tarski paradox.[k]

As for the role of the Axiom of Choice in their two decomposition theorems, the first for polyhedra in space and the second for polygons in the plane, Banach and Tarski were very explicit. They gave no encouragement to opponents of this axiom:

> One does not know how to establish either of these two theorems without relying on the Axiom of Choice—neither the first, which seems paradoxical, nor the second, which completely agrees with one's intuition. Furthermore, by analyzing their respective proofs, one can verify that the Axiom of Choice plays a much less substantial role in the paradoxical theorem than it does in the theorem agreeing with intuition [1924, p. 245].

Eventually, however, A. P. Morse [1949] established that the non-existence of paradoxical decompositions in the line and on the plane could be proved *without* this axiom.

In 1929 John von Neumann shed light on the Banach-Tarski paradox and, at the same time, proposed a more far-reaching generalization of the Measure Problem. It might appear, von Neumann observed, that the existence of a

[j]Banach, 1923, pp. 23–33. When he returned to these questions in his well-known book *The Theory of Linear Operators,* he derived his solution to the Extended Measure Problem from the Hahn-Banach Theorem [Banach, 1932, pp. 29–33]. It should be noted that this theorem too requires the Axiom of Choice.

[k]According to a personal communication in March 1982 from Tarski to the author, Banach and Tarski independently discovered the Banach-Tarski paradox, and then they decided to write a joint paper on the subject [1924]; however, the more general result on three-dimensional bounded sets with non-empty interiors was developed by Tarski alone.

suitable measure in one or two dimensions—as contrasted with the non-existence of such a measure in three or more dimensions—revealed a fundamental dichotomy between the nature of the Euclidean line or plane and that of Euclidean space. As he hastened to point out, this was not true. The essential difference did not arise from the dimensionality of the space but from the underlying group of motions. In order to clarify the matter, he revised Hausdorff's Extended Measure Problem by letting M be any set, W be any subset of M, and G be any group of one-one mappings of M onto itself. Then he defined m to be an (M, W, G)-measure if m assigned a non-negative real number to each subset A of M and if m had the following three properties:

(a) $m(W) = 1$.
(b) If f is in G, then $m(A) = m(f''A)$, where $f''A$ is the image of A under f.
(c) $m(A \cup B) = m(A) + m(B)$ if A and B are disjoint.

Here he altered Hausdorff's problem in two essential respects, first by replacing n-dimensional Euclidean space with an arbitrary set, and second by considering an arbitrary group instead of the congruence-preserving group of Euclidean motions. After he stated a sufficient condition (which involved Abelian factor groups) for an (M, W, G)-measure to exist, he proved that this condition is satisfied by the group of Euclidean motions in n-dimensional space for $n = 1$ and $n = 2$, but not for $n \geq 3$. On the other hand, he demonstrated that if G has a free subgroup with two generators and without fixed points (as was the case for the group of Euclidean motions in n-dimensional space whenever $n \geq 3$), then an (M, M, G)-measure does not exist. Finally, he showed that an (M, M, G)-measure fails to exist for the plane too if the group G is taken to be the affine group.[1] In all of these results, the Axiom of Choice played an essential role [1929, p. 73–87].

Later the same year, the Measure Problem was generalized in a different direction by Banach, who regarded the problem as essentially set-theoretic rather than geometric or even group-theoretic. By means of his generalization, he placed the problem (which he called the Generalized Measure Problem) squarely in the abstract theory of measure. This problem asked for a function m that assigns a non-negative real number to each subset A of the closed real interval $[0, 1]$ and satisfies the following conditions:

(I) $m(A) \neq 0$ for some A.
(II) $m(A) = 0$ if A contains exactly one element.
(III) m is countably additive.

Thereby Banach dispensed with the geometric requirement that congruent sets have equal measure and replaced it by the requirement that a point have

---

[1] Von Neumann also showed that a certain plausible choice of G, which, however, was not measure-preserving, led to a paradoxical decomposition of any line segment. During the summer of 1982 Stanley Wagon succeeded in showing that for the proper choice of G there fails to exist a (M, M, G)-measure on any line segment M. Wagon's proof was substantially simplified by Jan Mycielski during the fall (personal communication from S. Wagon).

measure zero. Without this latter requirement there would have been a trivial solution, namely, $m(A) = 1$ if A contains a certain predetermined point and $m(A) = 0$ otherwise. By weakening the geometric requirement of preserving congruence to an essentially set-theoretic condition, Banach was able to inquire if countable additivity could be reinstated.

In 1929 Banach published a joint article with another of Sierpiński's former students, Kazimierz Kuratowski. After Banach posed the Generalized Measure Problem, he found that he could solve it negatively if he could demonstrate a certain combinatorial lemma on countable unions. Banach and Kuratowski independently arrived at a proof of this lemma, by means of the Continuum Hypothesis, and so they published their results in a joint paper [1929]. Consequently, no measure of the desired sort existed if the Continuum Hypothesis was true.

Soon Banach took cognizance of the fact that the Generalized Measure Problem did not really concern the interval [0, 1] in particular but rather any set with the cardinality c of the continuum. As a result, one could pose the same problem for sets of arbitrary infinite cardinality. Thus in 1930 Banach introduced the Generalized Measure Problem for Abstract Sets: to find, for a given set E of cardinality $\aleph_\alpha$, a function m that assigns a non-negative real number to each subset A of E such that conditions (I), (II), and (III) hold. By using transfinite induction on α and by assuming the Generalized Continuum Hypothesis ($2^{\aleph_\beta} = \aleph_{\beta+1}$ for all ordinals β), he was able to show that the cardinal of the smallest E to have such a measure was weakly inaccessible [1930]. That is, the cardinal of E was uncountable, not the successor of any cardinal, and not the sum of a smaller number of smaller cardinals.

Meanwhile, Tarski had weakened Banach's Generalized Measure Problem for Abstract Sets by replacing condition (III) of countable additivity with condition (iv) of finite additivity. By means of the Axiom of Choice, Tarski showed that every infinite set E has such a measure and, furthermore, that the measure can be restricted to take only the values zero and one [1930]. In effect, Tarski obtained his result by constructing a maximal ideal on the family of subsets of E.

As Tarski acknowledged at the end of his paper [1930, p. 50], Kuratowski had pointed out to him that his result also followed easily from the argument in an earlier article by Stanisław Ulam [1929]. At that time Ulam was a twenty-year-old student at Lwow who, as a result of his paper of 1929, had decided to become a mathematician [1976, p. 30]. Encouraged by Kuratowski, Ulam attempted to improve Banach's result. Without assuming the Generalized Continuum Hypothesis or even the Continuum Hypothesis, Ulam showed that the cardinal of the smallest set E satisfying the Generalized Measure Problem for Abstract Sets must be weakly inaccessible and that the Generalized Measure Problem has no solution if there is no weakly inaccessible cardinal less than or equal to c [1930, p. 141]. As in the results of Banach and Kuratowski, the Axiom of Choice was essential here.

In more recent terminology, the cardinal of a set E satisfying the Generalized Measure Problem for Abstract Sets is said to be a real-valued measurable cardinal, whereas such a cardinal whose measure takes only the values zero and one is called a measurable cardinal. Thus Ulam established that the first measurable cardinal is strongly inaccessible (*i.e.,* weakly inaccessible and not the product of a smaller number of smaller sets) and that the first real-valued measurable cardinal is likewise strongly inaccessible if every weakly inaccessible cardinal is greater than c [1930, p. 150]. Independently, Tarski had shown that if a cardinal is non-measurable, then the cardinal of its power set is non-measurable [Ulam, 1930, p. 146].

Yet the question remained open as to whether there exists a measurable cardinal or even a real-valued measurable cardinal. Not until 1960 did Tarski (relying heavily on the work of his student William Hanf) establish that the first measurable cardinal κ, if it exists, must be extraordinarily large. In particular, there must be κ strongly inaccessible cardinals less than κ [Tarski, 1962].

CONCLUSION

In the absence of the Axiom of Choice, however, there were other possibilities. Indeed, it could then happen, as Thomas Jech demonstrated [1968], that aleph-one is a measurable cardinal—provided that there is a model of set theory with the Axiom of Choice in which there is a measurable cardinal. Moreover, Robert Solovay established that c may be a real-valued measurable cardinal since every set of real numbers can be Lebesgue-measurable [1965; 1970]. To obtain this result, Solovay had to postulate the existence of a strongly inaccessible cardinal—hardly an agreeable assumption to constructivists such as Baire, Borel, and Lebesgue.[m]

Thus Lebesgue's Measure Problem, which originated as a technical problem in mathematics, acquired philosophical ramifications—thanks to the Axiom of Choice. Subsequently, the philosophical problems existed side by side with new mathematical results. Lebesgue's philosophical scruples prevented him from pursuing fully the questions that his Measure Problem led others to raise. At the same time, his work reveals how a mathematician of the first rank may subtly fail to see how he is fundamentally violating his philosophical scruples in his own work. By contrast, Sierpiński and his Polish colleagues, without insisting that one accept the Axiom of Choice, explored its effects on the Measure Problem with clarity, depth, and consistency.

ACKNOWLEDGMENTS

A preliminary version of this paper was read at the University of Toronto on 22 March 1978. The present version was read at the New York Academy of Sciences on 25 March 1981, at Iowa State University on 9 February 1982, and

---

[m]The necessity of the inaccessible cardinal (provided that the Denumerable Axiom holds or at least that aleph-one is regular) is due to Saharon Shelah (1980). Solovay (1971) had also shown that, in set theory with the Axiom of Choice, the existence of a measurable cardinal is equiconsistent with the existence of a real-valued measurable cardinal.

at the University of Waterloo on 13 February 1982. I would like to thank Ivor Grattan-Guinness, Esther R. Phillips, and Stanley Wagon for their suggestions, both stylistic and substantive.

REFERENCES

BAIRE, R.
1898 Sur les fonctions discontinues qui se rattachent aux fonctions continues. Comptes Rendus Hebdomadaires des Séances de l'Académie des Sciences, Paris **129**: 1621–1623.
1899 Sur les fonctions de variables réelles. Annali di matematica pura ed applicata (3) **3**: 1–123.

BAIRE, R., *et alii*
1905 Cinq lettres sur la théorie des ensembles. Bulletin de la Société Mathématique de France **33**: 261–273; translated in Moore, 1982, pp. 311–320.

BANACH, S.
1922 Sur les opérations dans les ensembles abstraits et leur application aux équations intégrales. Fundamenta Mathematicae **3**: 133–181.
1923 Sur le problème de la mesure. *ibid.* **4**: 7–33.
1930 Über additive Massfunktionen in abstrakten Mengen. *ibid.* **15**: 97–101.
1932 Théorie des opérations linéaires. Monografje Matematyczne, vol. 1 (Warsaw: Garasiński); reprinted by Chelsea, New York, 1963.

BANACH, S. & K. KURATOWSKI
1929 Sur une généralisation du problème de la mesure. Fundamenta Mathematicae **14**: 127–131.

BANACH, S. & A. TARSKI
1924 Sur la décomposition des ensembles de points en parties respectivement congruentes. Fundamenta Mathematicae **6**: 244–277.

BETTAZZI, R.
1896 Gruppi finiti ed infiniti de enti. Accademia delle Scienze di Torino, Classe di Scienze Fisiche, Matematiche, e Naturale (Atti) **31**: 506–512.

BOREL, E.
1895 Sur quelques points de la théorie des fonctions. Annales scientifiques de l'Ecole Normale Supérieure **12**: 9–55.
1898 Leçons sur la théorie des fonctions. Gauthier-Villars. Paris.
1899 A propos de 'l'infini nouveau.' Revue philosophique de la France et de l'étranger **48**: 383–390.
1900 L'antinomie du transfini. *ibid.* **49**: 378–383.
1905 Quelques remarques sur les principes de la théorie des ensembles. Mathematische Annalen **60**: 194–195.
1912 Notice sur les travaux scientifiques de M. Emile Borel. Gauthier-Villars. Paris.
1914 Second edition of Leçons sur la théorie des fonctions, 1898.

CANTOR, G.
1871 Über trigonometrische Reihen. Mathematische Annalen **4**: 139–143.
1872 Über die Ausdehnung eines Satzes aus der Theorie der trigonometrischen Reihen. *ibid.* **5**: 123–132.
1874 Über eine Eigenschaft des Inbegriffes aller reellen algebraischen Zahlen. Journal für die reine und angewandte Mathematik (Crelle) **77**: 258–262.
1878 Ein Beitrag zur Mannigfaltigkeitslehre. *ibid.* **84**: 242–258.

1879 Über unendliche lineare Punktmannichfaltigkeiten, I. Mathematische Annalen **15:** 1–7.

1880 Part II of [1879]. *ibid.* **17:** 355–358.

1882 Part III of [1879]. *ibid.* **20:** 113–121.

1883 Part IV of [1879]. *ibid.* **21:** 51–58.

1883a Part V of [1879]. *ibid.* **21:** 545–586.

1883b Sur une propriété du système de tous les nombres algébriques réels. Acta Mathematica **2:** 305–310 (French translation of [1874]).

1883c Une contribution à la théorie des ensembles. *ibid.*: 311–328 (French translation of [1878]).

1883d Sur les séries trigonométriques. *ibid.*: 329–335 (French translation of [1871]).

1883e Extension d'un théorème de la théorie des séries trigonométriques. *ibid.*:336–348 (French translation of [1872]).

1883f Sur les ensembles infinis et linéaires de points, I. *ibid.*: 349–356 (French translation of [1879]).

1883g Part II of [1883f]. *ibid.*: 357–360 (French translation of [1880]).

1883h Part III of [1883f]. *ibid.*: 361–371 (French translation of [1882]).

1883i Part IV of [1883f]. *ibid.*: 372–380 (French translation of [1883]).

1883j Fondements d'une théorie générale des ensembles. *ibid.*: 381–408 (French translation of [1883a]).

1883k Sur divers théorèmes de la théorie des ensembles de points situés dans un espace continu à *n* dimensions. Première communication. Extrait d'une lettre addressée à l'éditeur. *ibid.*: 409–414.

COHEN, P. J.

1963 The Independence of the Continuum Hypothesis, I. Proceedings of the National Academy of Sciences USA **50:** 1143–1148.

COUTURAT, L.

1896 De l'infini mathématique. Alcan. Paris.

1896a Review of Hannequin, 1895. Revue de Métaphysique et de Morale **4:** 778–797.

1897 Continuation of 1896a. *ibid.* **5:** 87–113, 220–247.

DAUBEN, J. W.

1979 Georg Cantor: His Mathematics and Philosophy of the Infinite. Harvard University Press. Cambridge, MA.

DE LA VALLÉE-POUSSIN, C.

1934 Intégrales de Lebesgue, fonctions d'ensembles, classes de Baire, second edition. Gauthier-Villars. Paris.

DUGAC, P.

1976 Notes et documents sur la vie et l'oeuvre de René Baire. Archive for History of Exact Sciences **15:** 297–383.

EVELLIN, F.

1898 L'infini nouveau. Revue philosophique de la France et de l'étranger **45:** 113–119.

FEFERMAN, S. & A. LEVY

1963 Independence Results in Set Theory by Cohen's Method, II. Notices of the American Mathematical Society **10:** 593.

GRATTAN-GUINNESS, I.

1971 The Correspondence between Georg Cantor and Philip Jourdain. Jahresbericht der Deutschen Mathematiker-Vereinigung **73:** 111–130.

HANNEQUIN, A.
1895   Essai critique sur l'hypothèse des atomes dans la science contemporaine.
        Masson. Paris.
HAUSDORFF, F.
1914   Bemerkung über den Inhalt von Punktmengen. Mathematische Annalen
        75: 428–433.
1914a   Grundzüge der Mengenlehre. Veit. Leipzig.
HAWKINS, T.
1975   Lebesgue's Theory of Integration, second edition. Chelsea. New York.
JECH, T.
1968 ω_1 Can Be Measurable. Israel Journal of Mathematics **6**: 363–367.
JORDAN, C.
1892   Remarques sur les intégrales définies. Journal de mathématiques pures et
        appliquées (**4**) **8**: 69–99.
1893   Cours d'analyse de l'Ecole Polytechnique: Tome premier, Calcul différentiel,
        second edition. Gauthier-Villars. Paris.
JOURDAIN, P. E. B.
1904   On the Transfinite Cardinal Numbers of Well-Ordered Aggregates. Philosoph-
        ical Magazine (**6**) **7**: 61–75.
KERRY, B.
1885   Über G. Cantor's Mannigfaltigkeitsuntersuchungen. Vierteljahrsschrift für
        wissenschaftliche Philosophie **9**: 191–232.
KOWALEWSKI, G.
1950   Bestand und Wandel. Oldenbourg. Munich.
LEBESGUE, H.
1899   Sur les fonctions de plusieurs variables. Comptes Rendus Hebdomadaires des
        Séances de l'Académie des Sciences, Paris **128**: 811–813.
1899a   Sur la définition de l'aire d'une surface. *ibid.* **129**: 870–873.
1902   Intégrale, longueur, aire. Annali di matematica pura ed applicata (**3**) **7**: 231–
        359.
1904   Leçons sur l'intégration et la recherche des fonctions primitives. Gauthier-
        Villars. Paris.
1905   Sur les fonctions représentables analytiquement. Journal de mathématiques
        pures et appliquées. **60**: 139–216.
1907   Contribution à l'étude des correspondances de M. Zermelo. Bulletin de la
        Société Mathématique de France **35**: 202–212.
1907a   Sur les transformations ponctuelles, transformant les plans en plans, qu'on
        peut définir par des procédés analytiques. Accademia delle Scienze di
        Torino, Classe di Scienze Fisiche, Matematiche, e Naturale (Atti) **42**: 532–
        539.
1918   Remarques sur les théories de la mesure et de l'intégration. Annales scienti-
        fiques de l'Ecole Normale Supérieure (**3**) **35**: 191–250.
1922   Notice sur les travaux scientifiques. E. Privat. Toulouse.
1941   Les controverses sur la théorie des ensembles et la question des fondements. Les
        entretiens de Zurich, F. Gonseth, Ed.: 109–122. Leeman. Zurich.
LECHELAS, G.
1897   Review of Couturat, 1896. Revue de Métaphysique et de Morale **5**: 462–488,
        620–643.
LEVI, B.
1902   Intorno alla teoria degli aggregati. Istituto Lombardo di Scienze e Lettere,
        Rendiconti (**2**) **35**: 863–868.

MOORE, G. H.
1978   The Origins of Zermelo's Axiomatization of Set Theory. Journal of Philosophical Logic **7**: 307–329.
1982   Zermelo's Axiom of Choice: Its Origins, Development, and Influence. Springer. New York.
MORSE, A. P.
1949   Squares are normal. Fundamenta Mathematicae **36**: 35–39.
PEANO, G.
1890   Démonstration de l'intégrabilité des équations différentielles ordinaires. Mathematische Annalen **37**: 182–228.
SHELAH, S.
1980   Going to Canossa. Abstracts of the American Mathematical Society **1**: 630.
SIERPIŃSKI, W.
1916   Sur le rôle de l'axiome de M. Zermelo dans l'analyse moderne. Comptes Rendus Hebdomadaires des Séances de l'Académie des Sciences, Paris **163**: 688–691.
1917   Sur quelques problèmes qui impliquent des fonctions non-mesurables. *ibid.* **164**: 882–884.
1918   L'axiome de M. Zermelo et son rôle dans la théorie des ensembles et l'analyse. Bulletin de l'Académie des Sciences de Cracovie, Classe des Sciences Mathématiques, Série A: 97–152.
1941   L'axiome du choix et l'hypothèse du continu. Les entretiens de Zurich. F. Gonseth, Ed.: 125–143. Leeman. Zurich.
SOLOVAY, R.
1965   The Measure Problem. Notices of the American Mathematical Society **12**: 217.
1970   A Model of Set Theory in which Every Set of Reals Is Lebesgue Measurable. Annals of Mathematics **92**: 1–56.
1971   Real-Valued Measurable Cardinals. Axiomatic set theory. Proc. Symp. Pure Math. (Amer. Math. Soc.) **13**(1): 397–428.
STEPRĀNS, J.
1982   Some Results in Set Theory. Ph.D. dissertation, University of Toronto.
TANNERY, J.
1884   Review of Cantor, 1883b–k. Bulletin des sciences mathématiques et astronomiques (**2**) **2**: 162–171.
1886   Introduction à la théorie des fonctions d'une variable. Hermann. Paris.
1901   Notice sur les travaux scientifiques. Gauthier-Villars. Paris.
TANNERY, P.
1884   Review of Cantor, 1883b–k. Bulletin des sciences mathématiques et astronomiques (**2**) **2**: 162–171.
1885   Le concept scientifique du continu: Zénon d'Elée et Georg Cantor. Revue philosophique de la France et de l'étranger **20**: 385–410.
TARSKI, A.
1930   Une contribution à la théorie de la mesure. Fundamenta Mathematicae **15**: 42–50.
1962   Some Problems and Results Relevant to the Foundations of Set Theory. Logic, Methodology, and Philosophy of Science (Proceedings of the 1960 International Congress). Stanford University Press. Stanford, CA.: 125–135.
TONELLI, L.
1921   Fondamenti di calcolo delle variazioni, vol. 1. Zanichelli. Bologna.

ULAM, S.
1929   Concerning Functions of Sets. Fundamenta Mathematicae **14:** 231–233.
1930   Zur Masstheorie in der allgemeinen Mengenlehre. *ibid.* **16:** 140–150.
1976   Adventures of a Mathematician. Scribner's. New York.
VITALI, G.
1905   Sul problema della misura dei gruppi di punti di una retta. Tio. Gamberini e
       Parmeggiani. Bologna.
ZERMELO, E.
1904   Beweis, dass jede Menge wohlgeordnet werden kann (Aus einem an Herrn
       Hilbert gerichteten Briefe). Mathematische Annalen **59:** 514–516.

Reflections on Italian Medical Writings of the Fourteenth and Fifteenth Centuries[a]

NANCY G. SIRAISI

Department of History
Hunter College of the City University of New York
New York, New York 10021

Over the last twenty or thirty years, the study of all aspects of late medieval science and natural philosophy has progressed rather rapidly, aided by the appearance of new or expanded reference tools, critical editions of major texts, and numerous secondary studies. Within this general picture, the medical learning of the period has received a share—if not perhaps, in my opinion, always quite its fair share—of attention. True, by far the greater part of the large written output of medieval physicians is still available only in manuscript or in editions produced during the first century of printing. Nonetheless, thanks to the devoted bibliographical and monographic labors of a few scholars, investigators of the present generation can far more easily than any of their predecessors gain an awareness of the extent, richness, and complexity of the written record of medieval medicine.[b] Certainly, much work on currents of thought, patterns of debate, and regional and temporal variation will have to be done before any real historical synthesis can be achieved. But we are now in a better position than ever before not merely to identify, classify, and locate medieval medical writings, but also to frame the historical questions that should be asked of this material. I want here to illustrate this contention by suggesting some possible approaches to the category of medical writing with which I am personally most familiar, namely the Latin academic commentaries on translations of Greek and Arabic works, and the "disputed questions" on medical subjects produced in the environment of the north Italian universities between the late thirteenth and the fifteenth centuries.

It is probably still true to say that, except among a very few specialists, these scholastic productions have more often been the object of calumny than of study. On the one hand, relatively little of the recent and ongoing intensive re-examination of late medieval thought, which has made the neglect or downgrading of late scholastic expositions of theology and natural philosophy a thing of the past among serious scholars, has been focused upon medical texts. On the other, those who have interested themselves in thirteenth to fifteenth century medicine from the standpoint of the history of anatomy, physiology, and therapy regarded as cumulative bodies of knowledge, or from

---

[a]This paper was presented at the meeting of the Section of History, Philosophy and Ethical Issues of Science and Technology of The New York Academy of Sciences held on 19 November 1980.

[b]For a bibliographic survey of some of the recent work in the history of medieval medical learning, see Nancy G. Siraisi, "Some Recent Work on Western European Medical Learning, ca. 1200–ca. 1500," *History of Universities,* 2 (1982), pp. 225–238.

155

0077–8923/83/0412–0155 $01.75/2 © 1983, NYAS

that of professional or social history, have tended to focus not upon theoretical expositions such as are found in commentaries and questions, but upon aspects of education and practice which seem, rightly or wrongly, to be either innovative or rooted in the actual experience of practitioners.

Yet the reasons why the theoretical expositions of medicine produced between the thirteenth and the fifteenth centuries merit attention are rather obvious. In the first place, these writings constitute a body of discussion of scientific topics that is simply too large to be ignored in any serious attempt to come to terms with the nature of the scientific world of the late Middle Ages and early Renaissance.

Secondly, the influence of these works, and of the kind of medical education they represent, persisted into the crucial early modern period. The end of the fifteenth century, the terminus of the present paper, was certainly not the end of scholastic medicine in the Italian universities. True, certain cultural and technological changes which made their first appearance shortly before 1500—I am thinking, of course, of aesthetic values and artistic techniques permitting realistic anatomical and botanical illustration, of the interest of some humanists in the original Greek texts of ancient medical works, of printing—were ultimately to be of considerable importance for medical learning. These factors undoubtedly played a part in creating the pre-conditions for the shift in emphasis in medical education and the substantive advances in investigative method and anatomical and physiological knowledge that characterized the Italian universities in the sixteenth and early seventeenth centuries. Yet printed editions of a good many of the Latin commentaries on works of Hippocrates, Galen, and Avicenna written by Italian learned physicians between the late thirteenth and mid-fifteenth century were produced, presumably for the academic market, from about the 1470s until the mid-sixteenth century. Indeed, given the relatively small number of known manuscripts of certain of these commentaries, some of them may have achieved wider circulation after 1500 than before. We may further note that if printings of the late medieval and early Renaissance commentaries ceased after about the mid-sixteenth century, it was not because the commentary as a genre, or the works mandated in the curriculum and traditionally commented upon, had fallen out of favor. Instead, until the early seventeenth century, the old commentaries were replaced by new ones, produced at Bologna and Padua and printed at Venice and elsewhere.

In trying to understand more fully various medical commentaries and questions written in the university centers of Italy between the late thirteenth and the fifteenth century, and to weigh the place these productions should have in our picture of the medical science and education of the period, I have found myself formulating a few very simple and general propositions. These follow, along with some illustrative examples.

1. While learned medicine in the various university centers of western Europe had much in common, significant regional variations did exist. The study of particular groups of physicians, or particular centers or groups of centers, is therefore a worthwhile exercise.

2. A great deal of our knowledge of medieval medicine is based upon the study of a body of literature, namely the translated versions of Greek and Arabic medical works studied by, and the writings of, literate physicians. Although a substantial portion of this material purports to concern the practice of medicine, it is usually very unsatisfactory as a source of information about what was actually done by whom, to whom and with what results (it may be noted in passing, however, that valuable accounts of medicine in various medieval and Renaissance communities are now being constructed from archival sources). On the other hand, both works on *theoria* and works on *practica* convey copious information about medicine and the life sciences considered as a body of knowledge transmitted by the written word. In examining medical writings, therefore, there seems little reason to assign priority to the analysis of discussions of *practica* or to neglect theoretical expositions.

3. The content of scholastic medical commentaries and questions includes a dynamic element; certain topics aroused lively debate for a time, subsequently giving way to others, and so on. Historical understanding is likely to be improved by attempting to follow these discussions. (To say this is not, of course, to deny that much routine and wholly traditional pedagogy is also to be found in commentaries and questions; though it should be pointed out that even routine expositions must have served to acquaint generations of students with some fairly important issues—for example, the recognition and confrontation of the differences between Aristotelian and Galenic physiology, which became a standard topic.)

4. Medical learning was always distinct from, but also related to, other aspects of the scientific and, indeed, the philosophical thought of the period; it is advisable, therefore, to be aware of the probability of parallels and cross-currents between developments in medical thought and in natural philosophy.

Let us consider the first point, that of regional differentiation. It should be clearly understood from the start that the foundations of the body of learning transmitted by the north Italian faculties of medicine in the fourteenth and fifteenth centuries were in no way peculiar to them: the texts on which teaching was based were part of a common heritage derived by western Europe from classical antiquity and the Islamic world; much of the methodology of teaching and investigation—the formal structure of lectures on authoritative texts and disputations on problems arising from them, to which the production of the written commentaries and questions is closely related— paralleled that employed in a whole range of other disciplines in university centers all over Europe. The important differences that distinguish medicine in the north Italian universities from medicine in other centers of learning have to do, from an external standpoint, with the professional, social, and institutional context; and, from an internal standpoint, with the development of arguments carried on among Italian scholars. The point may become clearer if we briefly consider some of the sources of medical learning and teaching methods.

In the course of the thirteenth century, medicine appeared as a "higher" faculty (that is, as a subject taught to students who had already completed studies in liberal arts and natural philosophy) in most of the emerging universities of western Europe. Thus, for example, Bologna, which was already attracting medical students shortly after 1200, seems to have achieved the organization of a college of doctors of arts and medicine and probably also of a student university of arts and medicine some time between the 1260s and the 1280s (Sorbelli, 1940, pp. 105–128; Busacchi, 1948, pp. 128–144). Similarly, the modest beginnings of university organization of medical education at Padua probably date from around 1260 (Arnaldi, 1977, pp. 388–431; Siraisi, 1973, pp. 22–26, 143–147). But the study of medicine as a learned discipline in western Europe has a continuous history stretching back through the early Middle Ages to late antiquity; and the course of that history followed the general development of medieval Latin culture. The process of transmission of Hippocratic and Galenic medicine, like that of Aristotelian natural philosophy, broadened and deepened between the late eleventh and the early thirteenth century; in the same period, the incorporation into the western tradition of Arabic contributions to both disciplines also took place. Toward the end of the twelfth and in the early thirteenth century there occurred at Salerno and various transalpine centers of learning the formation and dissemination of the celebrated collection of classical medical texts known as the *ars medicinae* or *articella,* which was to remain a focus of study throughout the period with which we are concerned. The practice of writing Latin commentaries on some of those texts was taken up by Salernitan masters in the twelfth century (Kristeller, 1956, 1976).

Between the 1220s and the 1260s both physicians and natural philosophers became aware of, and started discussing, the differences between Aristotle's and Galen's understanding of the function of the heart and brain and the roles of the respective parents in conception, topics of debate that were to remain of central importance until the seventeenth century.

In the course of the thirteenth century, too, scholastic methodology, in the sense of the isolation of questions to be resolved by the application of Aristotelian logic to organize the results of an appeal to reason, authority, and (to a lesser extent) experience, became integral to the study of medicine as to other disciplines. Questions of this kind begin to appear in large numbers in medical commentaries, and more rarely as independent items, by the second half of the thirteenth century. It should be noted, however, that the scholastic treatment of medical questions in the thirteenth century was itself an outgrowth of an earlier tradition of posing and responding to questions on scientific and medical topics which flourished in the twelfth century and has ancient sources (Lawn, 1963, 1979).

This earlier history serves to explain why the medical curricula of the north Italian universities when these arose were probably not substantially different (although no doubt there were variations in detail) from that of Montpellier, or even from those of such northern centers of medical learning as Paris and Oxford.

In important respects, then, the theoretical medical writings produced in the north Italian academic milieu between the late thirteenth and the fifteenth centuries are the product of intellectual developments that were by no means confined to Italy or even to the field of learned medicine. How indeed could this be otherwise, when the learned physicians of Bologna, Padua, Perugia, and Siena studied medical and philosophical texts that were mostly part of western Europe's common heritage from Greek antiquity and the Islamic world, and when the organization of teaching in their schools in many respects paralleled that in other disciplines and other *studia*? Moreover, the ready penetration of the schools of northern Italy by intellectual influences emanating from other centers of learning (whether Salerno or the schools of northern Europe) has been abundantly documented. Indeed, I shall shortly be locating the topic of one *quaestio* within a larger complex of ideas that interested not only contemporary physicians elsewhere (at Montpellier), but also theologians; and pointing to the influence on the author of an Oxford philosopher. Yet, simultaneously, the medical works written in Italy seem to me also to reflect both a distinctive intellectual environment and the activities of a rather tightly knit regional community of scholars.

It is doubtless true to say that in the university centers of northern Italy and at Montpellier, medicine early achieved a position of more importance in the life of the schools than it was accorded elsewhere in Europe. In Italy a prosperous and unusually well-organized medical professoriate, which already in the thirteenth century constituted the nucleus of a largely secular learned profession, enjoyed considerable social and political influence (Bullough, 1966; Siraisi, 1981, pp. 25–71). Furthermore, the teaching of medicine in intimate institutional association with arts and natural philosophy assured for medicine a place in the mainstream of philosophical and scientific discourse. It was, no doubt, this favorable social and intellectual setting that enabled the learned physicians of Bologna in the years before and after 1300 *both* to introduce such deservedly celebrated practical innovations as the teaching of anatomy on the cadaver and the compilation of accounts of individual cases *and* to become notoriously prolific in the production of works commenting on medical texts in the light of the prevailing system of logic and natural philosophy.

As for the Italian schools as compared with Montpellier, there are indeed many parallels in development, but there are also significant differences. In the first place, Bologna in the first half of the fourteenth century and Padua at different times during the fourteenth and fifteenth centuries were more important centers of philosophical learning than was Montpellier. Thus, it seems probable that learned physicians in the former cities usually had more immediate access to debate over general philosophical and scientific issues than did their colleagues at Montpellier. Secondly, if the pattern of production of commentaries is any guide, there is some evidence to suggest that the Italian schools continued to enjoy vigorous life in the later fourteenth and fifteenth centuries, whereas the output from Montpellier seems to have diminished after about 1330. Thus, if one looks at the catalogue to be found in

the "Hippocrates Latinus" of surviving commentaries on two of the Hippocratic works that were most widely studied in the Middle Ages, namely the
Aphorisms and *Regimen in Acute Diseases* (Kibre, 1975, pp. 118–123; 1977,
pp. 259–278), one finds seven by authors associated with Montpellier, five of
them written between approximately the 1280s and the 1330s. Twenty-six
commentaries on these two works are attributed to known Italian university
authors, and they are pretty evenly distributed across two centuries: eight
belong to the fifty or so years ending in 1330; eight to the middle or later
fourteenth century; and ten to the fifteenth. (Late medieval decline is often
postulated for the *studium* of Bologna; but one suspects that its extent may
have been overestimated.)

But the chief reason for looking at the output of fourteenth- and
fifteenth-century Italian learned physicians as a distinct body of material is
the intense awareness of regional scholarship manifested by those physicians
themselves. Thus, the compilers of the Statutes of the University of Arts and
Medicine of Padua issued in 1465 ordered that in studying the *Microtechne* of
Galen, scholars should apply themselves either to the commentary of Torrigiano de' Torrigiani, a Florentine medical author educated at Bologna who
died about 1319, or to that of Giacomo da Forlì, who had been a professor of
medicine at Padua and other Italian centers of learning until his death in 1414
(Statuta ..., n.d. fol. XXIV^v). Furthermore, the Latin commentaries on
works of Hippocrates and Galen written at Salerno, Montpellier, and
elsewhere may have been studied but were relatively seldom cited by north
Italian medical authors. Instead, the latter carefully cited and discussed the
views of their own colleagues and recent predecessors. It is thus quite usual in
works in which few other medieval Latin authors except perhaps Albertus
Magnus are mentioned by name to find repeated references to the opinions of
such medical luminaries as Taddeo Alderotti, a leading professor of medicine
at Bologna who died in 1295, Pietro d'Abano, professor of medicine, philosophy, and astrology at Padua until his death in about 1316, Dino del Garbo,
who taught at Bologna, Padua, and Siena, and died in 1327, and so on. In
1342, Gentile da Foligno, an author to whom I shall return shortly, was able to
distinguish the opinions on a particular topic of eleven learned physicians who
flourished at Bologna and Padua in the half century before he wrote (Gentile,
1520, fols. 63r–65v). Other instances could be adduced, including the
notorious one of Tommaso del Garbo, son of Dino, whose repeated references
to the superiority of his father's views may perhaps be suspected of being
primarily designed to foster a family practice rather than to forward scientific
discussion (e.g., Tommaso, 1506, fol. lr: "Thome filii Dini itidem doctoris
acutissimi . . . summa").

Various factors no doubt contributed to the development of this regional
focus in north Italian medical scholarship. Among them was probably the
superior availability of manuscripts of locally produced works and, in the case
of certain authors who were contemporaries, an element of personal and
professional rivalry. No doubt, too, the fairly constant circulation of faculty
that seems to have taken place among the various north Italian *studia* also

played a part; in some instances, this mobility was no doubt due to the financial or other instability of certain *studia,* but it also appears to have been the case that the more successful (success being here measured by prolixity of authorship, the ability to secure senior teaching positions, and fame among contemporaries) professors of medicine moved frequently from place to place, and often increased both reputation and salary by so doing. While this state of affairs makes it hard to distinguish, for example, a "school of Padua," it would appear to legitimize the study of north Italian university medicine as a whole as a distinct intellectual endeavor. Within it, certainly, we will find different foci: we may look for patterns of academic interest, such as the effort to extend the range of original works of Galen studied in the schools that seems to have preoccupied some Bologna masters of the late thirteenth and early fourteenth century (Siraisi, 1981, pp. 100–105); for networks of masters and pupils; for scholarly debates pursued in writing across several generations. But all of these are to be found within what is essentially a single intellectual, professional, and social milieu. None of the foregoing should, however, be taken to suggest isolation from the mainstream of European thought: on the contrary, it is becoming increasingly evident that masters in the Italian schools rapidly became *au courant* with philosophical developments in the transalpine universities, and these inevitably produced repercussions in writings on medical theory.

Let us turn now to the second point, that is the place of commentaries and questions among the whole range of written works that constitute our record of late medieval and Renaissance medical learning. To disentangle the role and mutual influence of theory, pedagogy, and practice in the activities and writings of academically trained physicians of this period is no easy task—and not one that I propose to undertake here. But a few considerations about different varieties of written work fairly readily present themselves. Commentaries and questions are, of course, only one of a number of genres of academic medical writing. Quantitatively, the greatest body of written material pertaining to the study of medicine probably consists of the translated texts of the Greek and Arabic treatises that were studied. Partial and complete manuscript copies of some of the Hippocratic writings that formed part of the standard medical curriculum of the *articella* greatly exceed in number those of all known Latin commentaries thereon combined (Kibre, 1975, pp. 103–123; 1976, pp. 262–292; 1977, pp. 254–279 for *On Regimen in Acute Diseases* and the *Aphorisms*), and the same is probably true of the *Microtechne* of Galen and the *Canon* of Avicenna, two of the other works most frequently commented upon. Furthermore, although the production of a commentary on one of these treatises was clearly a likely road to success for an aspiring professor of medicine, it was not the only one. A survey, based on Thorndike and Kibre (1963), of the bibliography of fifteen fairly prolific and widely known Italian medical authors, the earliest of whom died in 1295 and the latest in 1473, yields three who apparently wrote no commentaries on medical works at all, or at any rate none known to have survived. All three were professors of medicine of some renown at Padua (one of the three,

Giovanni de' Dondi is admittedly best remembered for his achievements in the then related field of astronomy rather than in medicine).[c]

Other contemporary genres of medical writing included collections of medical recipes, various treatises with the word *practica* in the title, and *consilia* for individual patients, as well as such innovations, in form if not always in content, as Simone of Genoa's medical dictionary (printed Venice, 1486, etc.; Klebs 920.4), Michele Savonarola's vernacular treatise on child-rearing addressed to mothers (Demaitre, 1977) and the plague treatises that began to appear after the great epidemic of the mid-fourteenth century. Certainly, both Simone's reference work and Savonarola's expert advice on pediatrics for the layman (or rather laywoman) are early examples of a type of medical writing that was to have a long future. It is perhaps less clear that they were regarded as especially significant in their own day. As for *consilia* and treatises on *practica* there seems to be a good deal of evidence to suggest that, at least as far as the Italian scene is concerned, they incorporated a significant academic element. Most *consilia* are very far from being a complete or direct record of the treatment of a patient by a physician (and are quite often simply a written response to a request for advice from another physician about a case the consultant had never seen). Collections of *consilia* were deliberately compiled, I think carefully edited, and used for study. Works on *practica* seem in many cases to have been intended primarily to aid in teaching a branch of the university medical curriculum bearing this title. Lest I be misinterpreted, let me hasten to add that there is abundant evidence that even the most learned and academic Italian university physicians *engaged* in practice and spent a good deal of their classroom time transmitting information about disease and treatment. I mean merely that, at any rate as regards thirteenth to fifteenth-century Italy, most of the surviving Latin works on the subject belong to the same academic milieu (and often to the same authors) as the commentaries, and that it is easy to overstate the difference in scientific approach and content between the commentaries and some of the other works. Even distinction as to genre becomes difficult when one is faced with highly scholastic commentaries upon works which themselves consist chiefly of descriptions of illness, injuries, and treatment—for example the surgical portions of the *Canon* of Avicenna, treated by Dino del Garbo with full apparatus of questions, *dubia,* and Aristotelian natural philosophy (Dino, 1489). Perhaps we need to consider this entire body of writing as a fairly well-integrated whole, and to give commentaries and questions if not the pride of place that their authors, and contemporaries, ascribed to them, at any rate a full measure of our attention.

---

[c]The authors surveyed were, in addition to Giovanni Dondi (d. 1389), Taddeo Alderotti (d. 1295), Pietro d'Abano (d. ca. 1316), Dino del Garbo (d. 1327), Gentile da Foligno (d. 1348), Marsiglio di Santa Sofia (d. 1405), Pietro da Tossignano (d. 1410), Nicolo Falcuccio (d. 1411 or 1412), Giacomo (Jacopo) da Forlì, d. 1414, Matteolo of Perugia (fl. 1437), Ugo Benzi (d. 1439), Antonio Cermisone (d. 1441), Cristoforo Barzizza (d. 1445), Michele Savonarola (d. 1466), and Sigismondo Polcastro (d. 1473). Savonarola and Polcastro are not indicated as the authors of commentaries.

To illustrate my third and fourth points, the existence of a dynamic element in scholastic writings on medical theory, and the currents running between fourteenth- and fifteenth-century treatments of medicine and natural philosophy, I would like to present as an example part of a single fourteenth-century question. The author is Gentile da Foligno (who is said to have died of the plague in 1348). The subject of his question was "the reduction of medicine to act," that is the way in which powers of medicines are released to work healing, cooling, drying or humidifying effects in the human body (Gentile, 1520).[d] The question preoccupied a number of Italian physicians from about the end of the thirteenth century whose views were recorded and criticized by Gentile. It was in fact one of several lines of inquiry pursued by physicians when they came to consider in detail the implications of the Galenic idea of "temperament" or "complexion," that is the balance of the elemental qualities of hot, wet, cold, and dry in living beings. According to this theory, every plant, animal, and among men individual, and every organ of the human body had its own distinct temperament, or balance of the four qualities in varying degrees of intensity. Sickness occurred when the balance of qualities in an individual was thrown out of order; cure would follow when the patient took medicine in which the qualities were matched to his disordered complexion (*mala complexio*) in such a way as to restore its balance. Thus, the concept of qualities in medicine (found in varying degrees of intensity) was only one aspect of a large body of closely knit biological, physiological, and pharmacological theory (McVaugh, 1975b, pp. 3–136). This long-lived system of ideas was in various ways satisfactory and useful: it provided a broad conceptual framework and mode of explanation for a wide range of phenomena, and it fitted in with the general scientific principles contained in Aristotelian teaching about the elements. Yet when thirteenth- and fourteenth-century academic physicians embarked on detailed discussions of the ramifications of this theory as presented in the works of Galen, Avicenna, and other authoritative medical writers, and of its relation to various Aristotelian concepts, they soon became aware of areas of ambiguity, contradiction, and confusion, which it then became their task to resolve. One of the topics that early aroused particular interest was the way in which ingested substances were absorbed and transformed in the body—a problem that interested theologians and Aristotelian natural philosophers as well as physicians, since the process could be held to involve change in the form of the ingested substance and the matter of the ingesting body. Thus we find that aspects of this topic were discussed both by Thomas Aquinas (Hall, 1971, 11; Aquinas, ST 1, qu. 119) and by physicians who commented upon the chapter on ingested substances in the medical *Canon* of Avicenna (Taddeo Alderotti, in MS Vat.palat.lat. 1246, fols. 78v–97v, for example). Another area of interest

---

[d]This *quaestio* is briefly discussed in Charles H. Talbot, "Medicine" in David C. Lindberg, ed., *Science in the Middle Ages* (Chicago, 1979), p. 405, and is noted in Lynn Thorndike, "A Medical Manuscript of the Fourteenth Century," *Journal of the History of Medicine and Allied Sciences,* 10 (1955), 395, note 28.

was the identification and effects of different degrees of intensity in medicinal qualities, a subject to which as Michael McVaugh has shown us (1969;1975a), the physicians of Montpellier devoted much attention. Yet a third was Gentile's topic, namely "the reduction of medicine to act."

As Gentile explained, the problem arose because uningested medicines and medical ingredients are not as a rule perceptibly endowed with the qualities attributed to them: pepper (hot) and opium (cold) may both feel the same temperature on the hand. In *De complexionibus* (*De temperamentis*) 3.2 and elsewhere, Galen had accounted for this by stating that medicinal qualities are only potential until activated by the heat of the human body, but this formulation, critically examined, raised as many questions as it solved. In particular, how can medicines have different qualities if they all depend on the single quality of the body's heat for their actualization? How is it possible for the heat of the human body to activate medicines that are hotter than the body? How can cold medicines be activated by heat? These were the questions that sparked the protracted discussions summarized by Gentile of the way in which the change in medicinal qualities from potential to actual came about.

The proposed solutions recorded by Gentile ranged from the simple to the highly complex. A theory attributed (Gentile thought perhaps mistakenly) to Taddeo Alderotti, held that all medicines were made up of both hot and cold parts, hot parts being more numerous in hot medicines and cold ones in cold. The hot parts heated, the cold ones chilled, and only the numerical predominance of one or the other determined whether the overall effect of the medicine was heating or cooling. In this scheme, the function of the heat of the body in releasing the action of medicines was merely to divide the hot and cold parts from each other (Gentile, 1520, fol. 63r). By contrast, a much more elaborate and philosophically oriented explanation was provided by Pietro d'Abano, (Gentile, 1520, fols. 63v–64r; compare Pietro, 1489, fols. 192v–193v). Pietro referred his readers to Aristotle's discussion in the first book of *De generatione et corruptione* of the need for agent and patient to be alike in some respects and dissimilar in others. The gist of Pietro's explanation was that cold medicines could be activated by the heat of the body because contraries acted upon one another. Pietro provided a series of what he considered to be other instances in nature of a quality being activated, enhanced, or engendered by its contrary: for example, the heat given off by lime when slaked with cold water. Gentile pointed out that if Pietro's solution is correct for cold medicines, it then becomes necessary to assume that hot medicines are activated in a different way from cold ones; and that Pietro seemed to have abandoned or at any rate greatly modified another Aristotelian principle, namely that like begets like. Furthermore, Gentile denied that some of the examples adduced by Pietro were in fact instances of the generation of opposite qualities in nature. Others might be valid in themselves—for example, the generation by a horse of a mule which was both like and unlike its parent—but were not really analogous to the activation of cold medicine. Furthermore, it was difficult to understand precisely how Pietro thought the heat of the body stimulated coldness (by concentration of existing coldness? by creation of more coldness than existed before?).

Torrigiano de'Torrigiani, following Avicenna, emphasized that "hot" and "cold" are relative, not absolute concepts (Gentile, 1520, fols. 64r–v; compare Torrigiano, fols. 100r–101r). He pointed out that the presence of the quality of hotness can only be identified by comparison with something else that is less hot. It is always necessary to specify with what the comparison is being made—whether with another species, with the norm within a species (choleric, melancholic, etc. temperaments in men being defined by the extent to which they varied from the norm or ideal for mankind), or with another individual. As far as medicines are concerned, they can be described as hot only if they produce heating effects on a particular species, complexionate type, or individual. A medicine that heats a horse is hot in respect to a horse—but may not be so for an animal of another kind. Physicians should therefore inquire about the effects of a medicine rather than about its innate qualities. Furthermore, since one agent may produce different effects on different substances (fire, for example, blackens some things and whitens others), the heat of the human body may release a power of cooling in one medication and a power of heating in another. Again, the same medicine may produce very different effects in different individuals, but the physician can only find out these variations empirically.

Torrigiano's treatment—and, indeed, that in Avicenna's *Canon*—might well produce in the mind of a reader some doubts as to the tenability of the Galenic assumption that hot and cold were two separate absolute qualities (existing in a scalar series of distinct degrees of intensity) and some awareness of the limitations of complexion theory as a guide to therapy. However, Gentile's own exposition of Galen's views subjects them to more radical and explicit criticism. It is clear, moreover, that Gentile was familiar with very recent philosophical discussions about the latitude, intensity, and remission of forms. Gentile stated that Galen himself had held that hot and cold medicines were not activated in the same way. According to Gentile, Galen's theory was that hot medicines received the heat through which they acted from the human bodies to which they were administered; cold medicines acted through their own inherent coldness, receiving from the heat of the human body merely either breaking down (*divisio*) or rarefaction (*subtillatio*) and distribution to the different parts of the body. Gentile then set against this opinion the belief of some scholars, among whom he named the Oxford master Walter Burley (d. ca. 1345), that cold could have no separate form of its own, since it was merely a diminution of heat, and that cold and heat were the same *species* (Gentile, 1520, fol. 66r; for Burley's views, Maier, 1968, pp. 314–314). Moreover, Gentile next proceeded to dismiss three separate solutions to this objection as, respectively, "reprehensible," "confusion and burying of understanding in words," and as an embarrassing error on the part of an otherwise respected master. Of the second of these solutions, one proposed by Dino del Garbo, Gentile remarked "from this it appears that Dino was intellectually convinced [that is, of the truth of the opinion that hot and cold medicines cannot be activated in different ways], but out of reverence for Galen wished to dismiss the opinion" (Gentile, 1520, fol. 66ᵛ). Since, as noted, Gentile's question was written in 1342 (Thorndike and Kibre, 1963, col. 90), his remarks reflect a rapid absorption in the Italian medical schools of the medical implications of current philosophical and scientific discussions originating in northern European centers of learning.*

Space does not permit an examination here of Gentile's account of the various other shadings of early fourteenth-century opinion on the reduction of medicine to act; nor is it possible to explore the three solutions of the problem to which he was prepared to allow some validity, all of them based in large part on the views of Avicenna and/or Averroes; nor to follow the subject in the writings of such later fourteenth-century Italian learned physicians as Tommaso del Garbo, who devoted an entire treatise to the topic, while criticizing Gentile for his prolixity. All of these, I hope to treat more entensively at some future date. For the present, however, let me conclude with some general reflections on the nature of discussions of medical theory such as those reported in Gentile's question.

By the beginning of the fourteenth century, learned physicians had already achieved the assimilation of an important body of ancient texts and teaching; the establishment of medicine as a fully learned and in Italy highly prestigious university discipline, without abandoning its technical and craft aspects; and a focus on differences between ancient authorities (and inconsistencies within them) that drew attention not only to problems of some importance in themselves, but also to the existence of discrepancies within the body of ancient scientific learning. Like their colleagues in natural philosophy, they thereafter pursued a continued and minute analysis of ancient texts and concepts. As Gentile's exposition shows, we may expect to find in their writings occasional important insights and some significant criticisms of ancient authorities, although not any radical or wholesale departure from established procedures and interests. Edward Grant's characterization (1978) of scholastic natural philosophy seems also to apply to medicine: reliance on the question method certainly fostered the development and clarification of differences of opinion and the critical analysis of ancient authorities; but it also meant that the content of scientific works tended to be broken down into a series of separate problems that were relatively seldom considered in relation to one another. Thus, just as the effective criticisms of certain aspects of Aristotle's theory of motion advanced by some fourteenth-century natural philosophers by no means entailed any general repudiation of Aristotelian natural science on their part, so Gentile da Foligno and other scholastic physicians were able to raise fairly bold questions—that examined above is only one of a number of possible examples—about particular aspects of the idea of complexion, without in any way jeopardizing their adherence to the concept in general or to other Galenic principles.

The works on medical theory of scholastic physicians of the thirteenth to the fifteenth century constitute a significant body of medieval scientific and technical writing. As we have seen from Gentile's *quaestio,* the subject matter

*Interest in physical problems relating to heat seems to have been fairly widespread among late medieval Italian learned physicians; see Marshall Clagett, *Giovanni Marheni and Late Medieval Physics* (New York, 1941).

of this literature is not confined to medicine as such, but also embraces fundamental considerations about the life sciences. These works are simultaneously a chapter in the long history of the reception, comprehension, and criticism of the scientific legacy of the ancient world and the principal source of information about major areas of the scientific culture of their own age. Their further study is likely to be a rewarding enterprise for historians of medieval and Renaissance science and thought for some time to come.

REFERENCES

ARNALDI, G. 1977. Le origini dello studio di Padova dalla migrazione universitaria del 1222 alla fine del periodo ezzeliniano. La Cultura:Rivista di filosofia, letteratura e storia **15**: 388–431.

BULLOUGH, V. L. 1966. The Development of Medicine as a Profession: The Contribution of the Medieval University to Modern Medicine. Basel, S. Karger.

BUSACCHI, V. 1948. I primordi dell'insegnamento medico a Bologna. Rivista di storia delle scienze mediche e naturali **39**: 128–144.

DEMAITRE, L. E. 1977. The Idea of Childhood and Child Care in Medical Writings of the Middle Ages. The Journal of Psychohistory **4**: 461–490.

DINO DEL GARBO. 1489. Clarissimi artium et medicine doctoris magistri Dini de Florentia expositio super 3ᵃ et 4ᵃ fen [IV] Avicenne et super parte quinte . . . Ferrara, (Klebs 336.1). The title given here is drawn from fol. lr of the copy of this edition at the New York Academy of Medicine, in the absence of a title page.

GALEN. 1976. Burgundio of Pisa's Translation of Galen's Peri Kraseon, De complexionibus. R. J. Durling, Ed. Galenus Latinus I. Ars medica Abteilungen 2, griechlatein. Medizin, Bd.6,1. Berlin.

GENTILE DA FOLIGNO. 1520. *Questiones et tractatus extravagantes clarissimi Domini Gentilis de Fulgineo . . .* Venice. Questio 46, "Utrum medicina que dicuntur tales in potentia reducantur ad actum a caliditate nostri corporis," fols. 63r–71v.

GRANT, E. 1978. Aristotelianism and the Longevity of the Medieval World View. History of Science **16**: 93–106.

HALL, T. S. 1971. Life, Death, and the Radical Moisture. Clio Medica **6**: 3–23.

KIBRE, P. 1975, 1976, 1977. Hippocrates Latinus: Repertorium of Hippocratic Writings in the Latin Middle Ages. Traditio **31**: 99–126; **32**: 257–92; **33**: 253–295. Subsequent issues bring the catalogue to completion in eight parts.

KRISTELLER, P. O. 1956. The School of Salerno: Its Development and Its Contribution to the History of Learning. *In* Studies in Renaissance Thought and Letters, pp. 495–551. Edizioni di Storia e Letteratura. Rome.

KRISTELLER, P. O. 1976. Bartholomaeus, Musandinus and Maurus of Salerno and Other Early Commentators of the "Articella," With a Tentative List of Texts and Manuscripts. Italia medioevale e umanistica **19**: 57–87.

LAWN, B. 1963. The Salernitan Questions: An Introduction to the History of Medieval and Renaissance Problem Literature. Oxford.

LAWN, B. 1979. The Prose Salernitan Questions. Auctores Britannici Medii Aevi V. London.

MCVAUGH, M. R. 1969. Quantified Medical Theory and Practice at Fourteenth-Century Montpellier. Bulletin of the History of Medicine **43**: 397–413.

MCVAUGH, M. R. 1975a. An Early Discussion of Medicinal Degrees at Montpellier by Henry of Winchester. Bulletin of the History of Medicine **49**: 57–71.

MCVAUGH, M. R. 1975b. Introduction to Arnald of Villanova, *Aphorismi de gradibus. In* Arnaldi de Villanova Opera Medica Omnia. L. Garcia-Ballester, J. A. Paniagua and M. R. McVaugh, Eds. **II:** 3–136. Granada-Barcelona.

MAIER, A. 1968. Zwei Grundprobleme der scholastischen Naturphilosophie. Edizioni di Storia e Litteratura, Rome.

PIETRO D'ABANO (PETRUS DE ABANO, PETRUS APONENSIS). 1496. Conciliator differentiarum philosophorum et precipue medicorum . . . Venice (Klebs 773.5). Differentia 140, fols. 192v–193v.

SIRAISI, N. G. 1973. Arts and Sciences at Padua: The Studium of Padua Before 1350. Pontifical Institute of Medieval Studies. Toronto.

SIRAISI, N. G. 1981. Taddeo Alderotti and His Pupils: Two Generations of Italian Medical Learning. Princeton University Press. Princeton, NJ.

SIRAISI, N. G. 1982. Some Recent Work on Western European Medical Learning, ca. 1200–ca. 1500. History of Universities **2:** 225–238.

Sorbelli, A. 1940. Storia della Università di Bologna. I. N. Zanichelli. Bologna.

Statuta Dominorum Artistarum Achademiae Patavinae. Venice, n.d. Hain 15015. This volume contains the statutes of the University of Arts and Medicine enacted in 1465, with additions of 1495.

TADDEO ALDEROTTI (TADEUS DE FLORENTIA). Commentary on Avicenna, *Canon* 1.2.2.1.15, inc. "Quod comeditur et bibitur, etc. In hoc capitulo intendit. . . ." Biblioteca Apostolica Vaticana, MS Vat. palat. lat. 1246, fols. 78v–97v.

TALBOT, C. H. 1978. Medicine. *In* Science in the Middle Ages. David C. Lindberg, Ed.: 391–428. University of Chicago Press. Chicago, IL.

THORNDIKE, L. 1955. A Medical Manuscript of the Fourteenth Century. Journal of the History of Medicine and Allied Sciences **10:** 395.

THORNDIKE and PEARL KIBRE. 1963. *A Catalogue of Incipits of Mediaeval Scientific Writings in Latin.* 2nd edition. Cambridge, Mass., The Mediaeval Academy of America.

TOMMASO DEL GARBO. 1506. . . . *Summa medicinalis.* . . . Venice.

TORRIGIANI, TORRIGIANO DE'. 1512. *Turisani monaci plusquam commentum in Microtegni Galieni,* 3.36, fols. 100r–101r. Venice.